AF698333

LIBRARY
NETWORK

LONG DISTANCE SERVICES

A Buyer's Guide

For a complete listing of the *Artech House Telecommunications Library*,
turn to the back of this book...

LONG DISTANCE SERVICES

A Buyer's Guide

Daniel D. Briere
TeleChoice, Inc.

Artech House
Boston • London

Library of Congress Cataloging-in-Publication Data

Briere, Daniel D.
Long distance services: a buyer's guide / Daniel D. Briere.
p. cm.
Includes bibliographical references.
Includes index.
ISBN 0-89006-439-3
1. Telephone--United States--Long distance--Handbooks, manuals, etc. 2. Telephone--United States--Long distance--Technological innovations. I. Title.
HE8817.B67 1990 90-42994
384.6'4--dc20 CIP

British Library Cataloguing in Publication Data

Briere, Daniel D.
Long distance services: a buyer's guide.
1. Long-distance communication services
I. Title
384.043

ISBN 0-89006-439-3

© 1990 Artech House, Inc.
685 Canton Street
Norwood, MA 02062

All rights reserved. Printed and bound in the United States of America. No part of this publication may be reproduced or utilized in any form or by any means, electronic or mechanical, including photocopying, recording, or by any information storage and retrieval system, without permission in writing from the publisher.

International Standard Book Number: 0-89006-439-3
Library of Congress Catalog Card Number: 90-42994

10 9 8 7 6 5 4 3 2 1

Table of Contents

Chapter 1

Introduction

After six years of intense competition under deregulation, the long distance industry shows signs of growing up. Long distance products are nearly generic, with previously distinct long distance services now integrated into one long migration path for service users. Pricing is cut-throat, with differences among providers' per minute costs sometimes measured in the hundredths of a cent. Mergers and acquisitions are consuming many industry participants, moving the industry dangerously close to an oligopoly.

The market itself has forced these fundamental changes in long distance services. These movements will shape the future planning environment for network managers in the next decade. Long distance is heading into a new era where price will not be the major determinant that it was in the past. The rules for buying long distance are changing. No longer are savings of 30 percent to 40 percent commonly possible just from opening one's long distance traffic to bid. Competition has pushed providers closer to their operating margins, giving the lower cost, facilities-based carriers the edge in the market.

Value-Added Services Gain Importance

In this new environment, carriers are striving to differentiate their core products from competitors' offerings. More and more, users recognize value-added enhanced services such as fax networks, videoteleconferencing, 700 recordings, and customized call detail reports as critical factors in picking one carrier over another.

Despite all of this, buying long distance services today is actually becoming easier than it ever has been in the past. Unlike several years ago when the number of carrier offerings for simple outbound voice service seemed endless, today it is unusual for a user to face more than a few options for long distance service within his or her particular level of monthly usage.

Increasing use of advanced network capabilities is making telecommunications a strategic corporate tool. Corporate telecommunications managers, now more than ever, need to be aware of everything they can do with their network; this book describes their options.

Long distance is not just WATS and 800 service any more. Public network communications has changed from the predominantly voice WATS and 800 communications services to a range of integrated voice, data, image and facsimile products. Long distance is WATS, 800, virtual networks, 700 services, 900 services, multi-purpose calling cards, switched data services, private lines, integrated voice, data and video, and a whole range of other things. Cross-service pricing plans, long term contracts, custom tariff plans, service aggregator buying groups – these are just some of the latest pricing trends that complicate the buying decision today. Buying long distance means evaluating all of the different feature and pricing options available to the end user.

It is hard to just keep up with all of the different ways that the same type of long distance service may be purchased. The crucial decisions of which carrier to choose is less of a pricing selection and more of a value selection – just how much can the user get for a dollar?

About this Book

This book attempts to introduce the telecommunications manager to the range of intrastate, interstate, and international long distance and enhanced services available today. With this knowledge as a foundation, network planners will be well prepared to attack network design from a global corporate perspective, not from a piecemeal applications-specific point-of-view. Indeed, the most effective networks today are those that are integrated, making most efficient solutions of common needs.

Importantly, this book is designed to give a manager a fresh look at the industry, and a view of the new technologies and services available in the marketplace. For present telecommunications managers, it provides a good overview of everything that he or she should be investigating. For new entrants, it provides a broad perspective of the industry. While this book does cover some noteworthy introductory topics, it is not designed to be a novice's guide to long distance. There are basic manuals for beginners which can provide the background necessary to later discussions. The intent was to keep the book readable for a knowledgeable audience in a three-hour sitting.

The remainder of the book provides a quick glimpse at the marketplace. Chapters 2 through 4 cover the basics of telecommunications that everyone needs to know. Most readers would be tempted to bypass these chapters – they should not. The market environment has changed substantially in recent years. No one knows everything, and the recent changes in the intelligent network have strong implications for nearly every long distance service. Also, most telecommunications professionals at one time or another have confessed they had an "embarrassingly simple" question that they were afraid to ask, because they were already expected to know the answer in the first place. Books such as this one let users find out without anyone knowing.

Chapter 5 discusses the most basic outbound service – message telecommunications service (MTS) – and its higher-end counterpart, wide area telecommunications service (WATS). MTS services do not really exist anymore and WATS services have changed dramatically. While low- to medium-sized customers have fewer services from which to choose, there are numerous optional features and pricing constructs that are critical for any user site with $8,000 or less in monthly long distance expenses.

Chapter 6 covers toll-free (800) services. Some of the most significant market growth in the past few years has come from 800 traffic. It is hard to find a print advertisement or television commercial without an 800 number listed. Intelligent network developments have enhanced 800 services to the point where inbound call centers double as strategic market research centers. In addition, the automation now available from the long distance 800 products is remarkable – alternate routing, automatic number identification, time-of-day and day-of-the-week routing, etc.

Chapter 7 deals with private lines. T-1 access and long haul circuits are commonplace now. T-3 is on its way, and switched T-1 data services are imminent. Private lines allow companies to gain further cost efficiencies on high-volume corporate routes. Despite great pricing and performance for dedicated facility-to-dedicated facility virtual network calls, there will always be networks that justify installing private facilities.

Chapter 8 discusses virtual networks – the next step up in the migration path for most WATS and 800 users. Virtual networks are the product of the future for long distance carriers. They combine the advantages of private line networks with the efficiency and flexibility of the public switched telephone network. Virtual network products are now capable of carrying both voice and data over switched facilities. In addition to the interexchange carriers, the Bell operating companies (BOCs) and foreign telecommunications carriers are launching complementary services, setting the stage for truly integrated private networks.

Chapter 9 involves calling cards. It is hard to decide how to name these plastic cards these days – network gateways might be more applicable. Among other things, these cards now can be used to call home from dozens of foreign countries, to buy goods though VISA or Master Card affinity arrangements, and to send and receive facsimile transmissions when traveling on the road. In the past, these calling cards were reserved for corporate management for use in isolated company situations. Today, these multi-use cards are integrated into company-wide virtual network plans, fax networks, and global calling schemes. They are an extension of the corporate network and deserve the attention devoted to them in this book.

Chapter 10 covers fax networks. Facsimile usage is rapidly growing within most corporations; yet facsimile management is often the farthest concept from most managers' minds. Fax networks are one of the first enhanced services that deal explicitly with this exploding market. While sometimes more expensive than direct usage costs, fax networks can add a great deal of functionality to existing machines.

Chapter 11 deals with 900 Services. "Paid-for" services will play an increasingly important role in future corporate networks. Today, they are used primarily for some certain questionable applications; however, solid corporate applications are starting evolve and 900 is gaining acceptance by the dialing public. Easy to use and easily accessible from any phone, 900 numbers are a key to future corporate information distribution.

Chapter 12 takes a look at cellular services – as stand-alone products and as access mechanisms for long distance services. The concept of a mobile office is catching on and low-cost supporting hardware is available – portable fax machines, data terminals, cellular phones, etc.

Chapter 13 examines billing and the various ways that a user can get a long distance call detail each month. The proliferation of electronic media into billing means a wide range of options for users.

Chapter 14 looks at the intrastate implications of corporate networks. As state public service commissions open local markets to competition, users have more options for originating and terminating long distance traffic. Metropolitan bypass and customer-originated network bypass are more commonplace today.

Chapter 15 extends the domestic US discussion overseas. Where separate international products existed in the past, domestic interstate offerings are now being modified to include overseas calling in a more integrated manner. Some of the oldest international network management practices are being redefined as the age of the global network approaches. Advances in international signaling and integrated services digital network (ISDN) capabilities are providing the platform for easier communications among nations. As foreign markets open – Japan and Europe in particular – domestic firms find themselves expanding overseas. This has important implications for all U.S. corporate telecom managers.

Chapter 16 delves into carrier pricing plans and present strategies for maintaining competitive pricing in a maturing market. Bargaining for the best deal is more important than ever. The average manager has numerous options for buying the same services. An AT&T customer, for instance, typically buys services from AT&T under FCC Tariffs No. 1, No. 2, No. 4, No. 9, No. 11, No. 12, No. 15, and No. 16. But the same service is often available from service aggregators, sales agents – even from AT&T's competitors.

Chapter 17 brings it all together and looks to the future. Where is long distance heading from here? This chapter talks about some of the futuristic developments that will be coming in the next decade.

The closing appendix includes an extensive glossary.

Some notes on terminology. The term "telecom manager" is used frequently throughout this book for convenience; it implicitly includes datacom manager, network engineer, or whatever term your organization applies to the decision-maker over your telecommunications network. Also, local exchange carrier is used to reference the provider of local telephone service in your area. This can be one of the Bell operating companies or one of the hundreds of independent carriers nationwide. While the emphasis of the discussion is on the BOCs, most also applies to independent telephone companies as well.

Finally, a few words of thanks to those who made this book possible. I give much appreciation to Mark Langner at TeleChoice for making the constant changes to the text of the book. His help is more appreciated than I often relate to him. Also, to all the product managers and press relations personnel at the carriers who helped edit the contents of this book, notably Pam Small at MCI and Mitch Montagna at AT&T. Finally, to Holly, my wife, for her undying support as I grapple with the never-ending requirements of telecommunications entrepreneurship.

Chapter 2

The Long Distance Marketplace

Keeping abreast of technical developments is among the most challenging aspects of the telecommunications industry. Technology, services, pricing, even the providers – everything is in a constant state of flux.

Telecommunications emerged as a reality from the local, national, and international desire of individuals and communities to exchange information and ideas. Thus, one finds global political, social and economic developments playing a key role in defining the growth of telecommunications products and services.

Telecommunications is a daily fact of life. When President Richard M. Nixon first visited China in 1971, television crews brought their own satellite communications terminals. When fires gutted Yosemite National Park in 1988, complex cellular telephone networks were established overnight to aid the effort. When the US invaded Panama in 1990, state-of-the-art battlefield-approved fiber optic and satellite communications gear went along.

There is no doubt that telecommunications has become pervasive throughout America, and the industrialized nations. The telecom manager's task is making sure that he or she is keeping up with the latest developments to best utilize what today's technology has to offer.

Looking Back

For most people, having to choose a long distance carrier came as quite a shock in the mid 1980s as changes in the industry rolled across the nation. For most of this

Figure 2.1 Before sophisticated switches were developed, each telephone had to be interconnected with each other telephone. The result was wire jungles like this one (Photo courtesy of US Sprint).

century, AT&T, affectionately called "Ma Bell," provided everything required for communications, including the telephone sets, local service and long distance transmission.

The first days of this century saw a form of telephone service quite different from today's. The telephone industry at the time was categorized by hundreds of small, local telephone operating companies that effectively wired everyone to everyone else in the surrounding community (see Figure 2.1). One by one, these firms were consolidated into larger telephone companies as the natural movement towards the inherent economic advantages of large scale operations started to take place.

AT&T was one of the companies which experienced this fast growth. By the time World War I began to make headlines around the globe, AT&T and its associated Bell telephone companies had garnered nearly 80 percent of the residential and business telephone customers in the United States. Its inter-city network was by far the largest in the nation and showed little signs of halting its national expansion.

The US government at the time was extremely fearful of monopolies, spurred primarily by developments in the oil industry. The Department of Justice confronted AT&T about anti-trust implications of its growth and forced the carrier to sign a document known as the Kingsbury Agreement. This accord limited AT&T to its operating area at that time and installed the means for independent, non-Bell telephone companies to interconnect with AT&T's inter-city network.

Still, industry growth favored AT&T in the marketplace – predominantly because of the vertical strength of its product lines. AT&T not only provided local and long distance service, but also provided and maintained all telephone equipment involved. Its Western Electric subsidiary manufactured all of the equipment required for transmission and switching – both in the home and office, as well as in the public network.

In 1949, the Justice Department again intervened, seeking to force AT&T to divest itself of at least the Western Electric subsidiary. Seven years later, in 1956, a consent decree was signed that allowed AT&T to retain ownership of Western Electric as long as the company only manufactured telecommunications equipment.

In the 1950s and 1960s, AT&T saw substantial increases in its traffic as the post-war prosperity overflowed into communications as well. During this period AT&T introduced some of its "new" services, such as 800 numbers.

As AT&T's market dominance became more alarming, complaints from competitors in the local, long distance and equipment markets prompted yet another suit against AT&T by the Department of Justice in 1975. For seven years, AT&T and the Department of Justice wrestled in the courts until AT&T finally agreed to restructuring in January of 1982.

This momentous agreement became known as the Modified Final Judgement or MFJ. It was approved by U.S. District Court Judge Harold Greene in August 1983 and became effective on January 1, 1984. It is this agreement that serves as the basis of today's competitive market.

The New AT&T

The resulting telecommunications marketplace was decidedly different. While AT&T was able to keep its interest in Western Electric, the company "divested" itself of 22 of its 24 local Bell operating companies (BOCs). In return for this, AT&T was freed from previous constraints that had kept AT&T out of non-telecom businesses, notably the computer industry. AT&T was allowed to keep its ownership interest in Southern New England Telephone (SNET) and Cincinnati Bell.

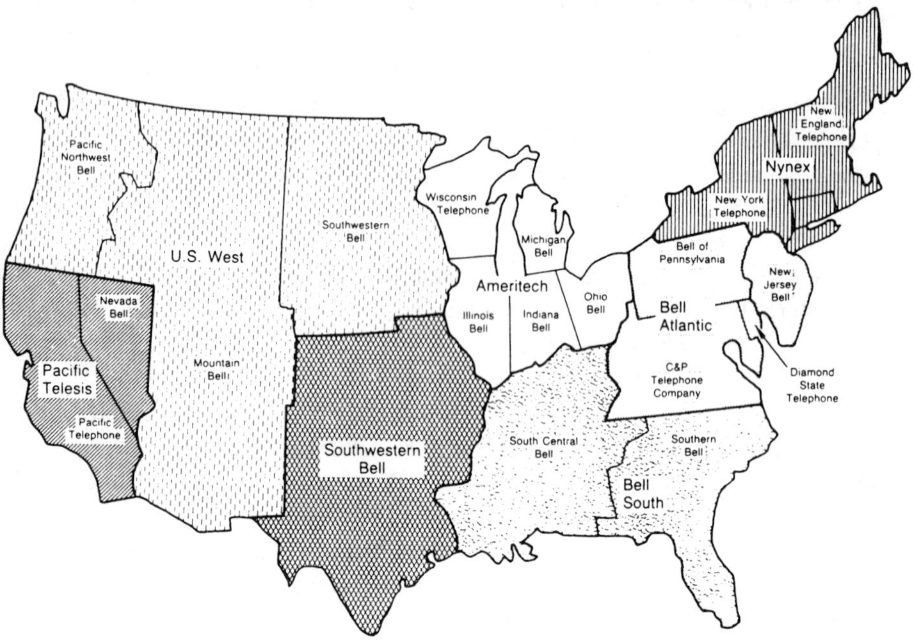

Figure 2.2 The seven regional Bell operating companies (from Elbert, B., Private Telecommunications Networks, © 1989, Artech House, Inc.).

At divestiture, there was enormous concern that the government was not taking the correct course of action. There was fear that, in separating AT&T's long distance operations from the local telephone operations, local telephone prices would soar. In years before, AT&T's long haul revenues had subsidized local telephone costs, providing a sort of "Robin Hood" equalization factor – those who could afford long distance calls subsidized local telephone service for the underprivileged.

Fortunately, this did not occur. Access charges paid by all subscribers and IXCs, plus PUC rate hikes, made up for the AT&T subsidies. In addition, service quality improved as equal access allowed all of the IXCs to offer dial 1 service. There were intense debates and numerous Federal Communications Commission (FCC) orders, Congressional bills, and court rulings safeguarding the divestiture process. Five years later, the industry is fairly agreed that these early actions protected the goal of affordable universal local telephone service.

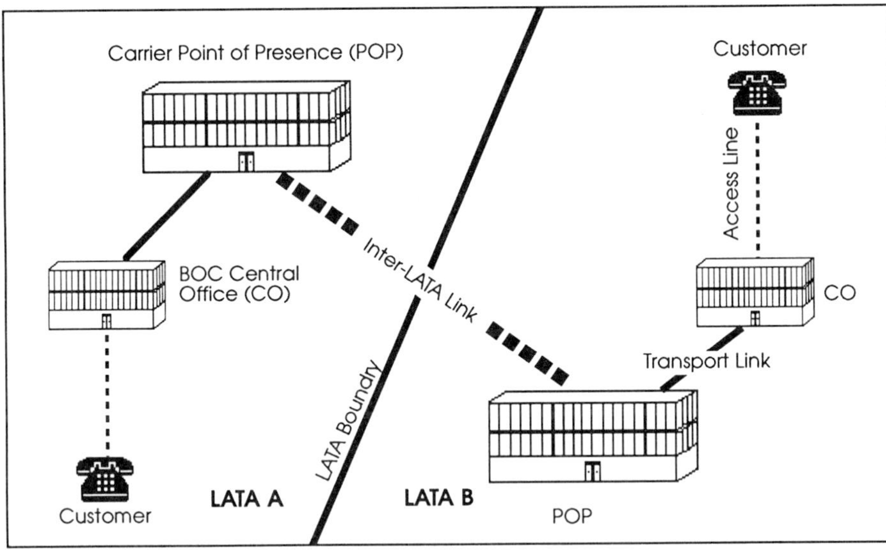

Figure 2.3 The path of an inter-LATA call.

The New RHCs

Certainly the most significant aspect of the divestiture process was the spin-off of the 22 Bell operating companies. These BOCs were organized into seven regional holding companies (RHCs), as shown in Figure 2.2.

With divestiture came territorial fights. Such confrontations had not arisen before because all of the groups were part of the same company – Ma Bell. Now, the individual Bell operating companies and AT&T were vying for pieces of the same pie. The lines of demarcation had to be drawn.

The BOCs wanted the courts to design their operating areas in accordance with state boundaries. To a degree, this made sense. Both AT&T and the BOCs already had existing organizational structures to deal with interstate and intrastate communications. Under this scenario, AT&T would get all of the interstate toll (long distance) traffic and the BOCs would get all intrastate toll traffic and be required to provide local area (non toll) calling.

Predictably, AT&T wanted the BOC operating areas to be defined in a much narrower manner – within the local service areas. This arrangement would give AT&T all intrastate and interstate toll traffic.

Compromise netted the local access and transport areas (LATAs). A LATA defines the operating territory of the local BOC. Strict rules were drawn barring BOCs from carrying any traffic outside or across LATA boundaries – this is reserved for the long distance, or interexchange, carriers (see Figure 2.3). LATAs coincide closely with standard metropolitan statistical areas (SMSAs) – government-categorized geographical areas that create communities of interest based on social, economic, and political factors.

There are 161 Bell LATAs and 28 independent marketing areas as of June 1, 1990. LATAs are added or altered according to changes in population or other demographic factors.

Only 40 percent of the nation's telephone exchanges are governed by Bell LATAs. Non-Bell areas are served by the more than 1400 independent telephone companies nationwide. Independent local telephone companies have operating areas known as independent marketing areas (IMAs).

The divestiture judgement and requirements do not generally apply to the independent telephone companies. Only GTE Corp, when it first entered the long distance market, was constrained in any manner. GTE had to sign a similar consent decree as the modified final judgement.

Equal Access

Judge Greene produced the second major fallout of divestiture through the imposition of equal access on the local telephone companies. Simply, equal access means that each local telephone company must give equal quality interconnection to all carriers requesting links to the local network.

In the post-divestiture environment, any domestic US long distance call typically requires local telephone company facilities on each end of the call. This places quite a bit of economic power in the hands of the local telephone companies. It would be extremely self-defeating to the whole divestiture process if a local carrier were to discriminate against the other carriers because of their long standing relationships with AT&T.

For this reason, equal access in the local networks was a critical cornerstone to achieving de facto competition in the long distance marketplace. Basic economic principles dictate that pure competitive environments can occur only when there is freedom of movement from one product to another. Therefore, it was critical to a free long-distance marketplace that a user could easily switch among interexchange carriers.

AT&T's advantage in terms of easy dialing, fast call connection times, and better quality connections hindered early long-distance competitors. Carriers such as MCI and GTE Sprint (now US Sprint) required users to dial up to 14 extra digits to complete a call. The only alternative was clumsy, unreliable autodialer equipment. AT&T on the other hand only required customers to dial 1 plus the telephone number to complete the call. This is commonly known as "one plus" dialing or "dial 1" access.

Additionally, the new competitor's customers also suffered poorer quality connections to the local network because the higher quality connections were simply not yet available. The local telephone companies had to upgrade equipment in each end office to make better interconnection available.

Judge Greene recognized these deficiencies and, as part of his on-going oversight of the divestiture, ordered the conversion of all Bell operating company end offices to dial "1" equal access for all carriers. As of June 1, 1990, the vast majority of all BOC and independent telephone company end offices had been converted to equal access. Only NYNEX, the BOC serving New York and New England, was lagging in conversion of its switches.

While equal access conversion is nearly complete, it has taken years. In the face of such competitive disadvantage, Judge Greene ordered the BOCs to provide the users of the lesser quality connections with lower rates. It was these lower access and egress rates that enabled many of AT&T's competitors to offer substantial discounts relative to AT&T's regular prices.

These origination and termination access costs represent a significant portion of any carrier's costs. In 1989, for instance, AT&T paid nearly $15 billion to the local telephone companies for this interconnection. That's nearly 50 percent of its revenues.

The Market Today

Today, headaches in local exchange connections still exist for AT&T's competitors, but they are far less disruptive than five years ago. The range of competing services has grown substantially and each of the major competitors to AT&T has a wide range of products.

Perhaps the first step in getting a grasp on today's market is finding out who the players are in the marketplace. Believe it or not, there are more than 1500 long distance companies presently providing service in the United States. Most people can name at best ten of these.

There are many different brands of long distance available on the market today from different groups that all provide fundamentally the same thing – long distance calling to the end user – but in crucially different ways. Some do nothing more than sell another carrier's telephone services. Others provide the billing and customer service functions, but not the transmission. Others provide everything you need – from your telephone to local telephone service to long distance lines.

In general, most end users have a limited choice of long distance carriers, determined by their physical location. Major metropolitan areas tend to have a greater number of carriers because there are more niches to exploit. Rural areas, by comparison, can have remarkably few options for service – sometimes just AT&T.

The Market Players

Most end users call any long distance provider a long distance "carrier." Carrier is a misnomer – primarily because some long distance providers never physically carry customer calls at all.

For definition's sake, let's agree to the following categorizations of long distance providers:

- **Carriers.** Carriers occupy the highest end of the service provider spectrum. These companies carry traffic over their own facilities and have truly national and international networks. While some of the transmission capacity is leased, the vast majority of it was installed by the individual company or acquired through various mergers. These providers have their own sales forces, customer service and accounting departments, and billing facilities. Carriers as a group usually will sell their services to resellers, rebillers, aggregators, and third party marketing groups for resale. Depending on the relationship the carrier has with each of the groups, customers sold long distance service may or may not remain a "customer" of the carrier. The better known carriers are AT&T, MCI and US Sprint.

- **Resellers.** These entities represent the second tier of the market. Resellers do precisely what their name implies – resell carrier services. Simply, they purchase bulk facilities from carriers for resale under their own names. Over their highest capacity routes, resellers will lease private circuits to achieve greater cost efficiencies. Resellers typically have their own switches, billing, sales, and customer service departments. Since the reseller remits monthly bills, the end user is a customer of and receives service from the reseller. Resellers range in size from very small regional companies with one switch, to vast national networks of mostly leased capacity. Cable & Wireless North America is surely one of the largest national resellers.

- **Rebillers (switchless resellers).** Similar to resellers, rebillers also purchase service from carriers. They receive master billing tapes from the carrier and rebill the service at their own stated rates. These rebillers, however, have no switch or network, but typically have sales, customer service and billing facilities. The end user is a customer of the rebiller but receives service from the carrier.

- **Service aggregators.** Similar to rebillers, service aggregators are entities which buy carrier services under multilocation volume discount plans. Then they enroll other businesses as sites with the intent of gathering volume discounts based on total calling at all sites. The end user is likely to remain a customer of the providing carrier, and is billed by the carrier with discounts based on all sites enrolled. Service aggregators do not own facilities and typically do not do billing for customers. They are marketing agents. For maintaining accounts, aggregators typically have customer service departments. Aggregators typically receive revenues from either monthly management fees or a negotiated split of "improved" or "incremental" savings. If service aggregators take on the billing function, they become rebillers.

- **Sub-aggregators.** Sub-aggregators are the same as aggregators, except they provide service off of another aggregator's discount plan due to their own limited financial resources or other limitations.

- **Agents.** These individuals or companies sell carrier services for the carrier on a commission basis. These companies differ from aggregator groups in that aggregators maintain a closer relationship with the customer after the sale is made. Agents make the sale and move to the next account – the end user becomes a customer of the carrier. Classic examples of these are seen daily in the interconnect industry, where there are many approved Centrex and cellular sales agents.

It is important to recognize the differences between each of these groups. They all sell something different. Accordingly, they offer varied advantages for each end user.

While the main focus of this book is on the major carriers, most of the services and issues described apply equally as well to resellers and rebillers.

The Major Carriers

More than 90 percent of all long distance in the US is sold by the top three major interexchange carriers – AT&T, MCI and US Sprint. At the middle of 1990, AT&T still retained 67 percent of the market, MCI had 13 percent and US Sprint had 8.5 percent. Early in 1990, MCI announced a commitment to acquire Telecom*USA, the fourth largest carrier. That acquisition would add approximately 1.5 percent to MCI's existing marketshare.

Each carrier network is in a different stage of development, partly due to the differing priorities set by upper management. In general, AT&T and MCI have sought to expand upon network intelligence while US Sprint has sought to develop its physical network. Each has excelled in their efforts. AT&T and MCI have greatly expanded their range of additional features and options, while US Sprint has arguably the best physical transmission network of the three.

Here are some specific characterizations of the largest carriers:

- AT&T is stressing its feature-rich network. The carrier has always had an abundance of capability inherent in its network and the use of its sophisticated network signaling system, CCS6. This functionality has been carried over in AT&T's switch to Signaling System 7 (SS7), an advanced network signaling protocol. AT&T has long been able to cross-utilize network developments across product categories, and not surprisingly, its virtual network service is largely a combination of various features and functions provided with other tariffed long distance services. Some of the network intelligence in place to provide advanced routing features for 800 services, for instance, can be readily applied to virtual network services as well. AT&T's weakness, however, has been its network plant. Much of it is outdated – there are still analog transmission facilities within the network. Late in 1988, AT&T decided to

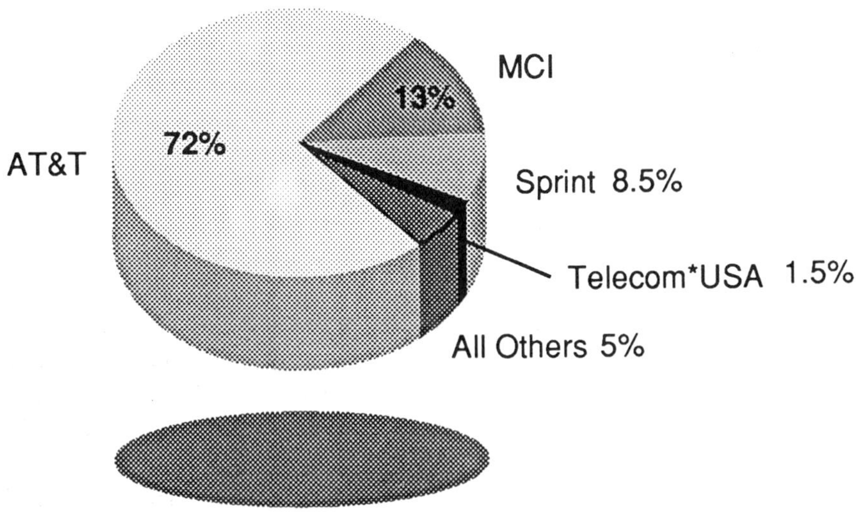

Figure 2.4 Long distance market share, 1990.

accelerate its network development and wrote off large portions of its network so that it could more quickly install digital facilities. In 1988, AT&T installed a software upgrade in its switches that included Digital Preference Switching. This that basically tells the switches to first choose digital facilities in routing a call. Other changes will continue through 1990, with a fully-digital network completed by the end of 1990, all analog private lines will be eliminated. AT&T is presently the market leader in the implementation of ISDN in its network.

- MCI has stressed its network management capabilities. In 1989, the company launched its integrated network management system, called INMS. This INMS package anchors MCI's intelligent network, providing the user with the ability to easily monitor, analyze and control any part of his or her network from a single terminal. MCI has sought to compete against AT&T feature-by-feature, product-by-product, and presently offers a full slate of service features that competes featurewise with AT&T's line, and which cost less. MCI's weaknesses also fall to its plant, which it is steadily upgrading. The carrier has completed Signaling System 7 implementation throughout its network, cutting call set-up times. However, nearly 25 percent of the network is still analog, covering approximately 10 percent of MCI's traffic. As a result, MCI created separate sub-networks, such as the Digital Data Network (DDN) to handle certain services requiring digital transmission. MCI says it will migrate to an all-digital network by 1993. The acquisition of Telecom*USA will help bolster MCI's fiber transmission capabilities in the South and Mid-West.

- US Sprint has emphasized primarily its physical plant, marketing itself as the only 100 percent fiber, all-digital network. US Sprint also uses a single switch architecture – all Northern DMS-250 or DMS-300 switches – allowing the carrier to achieve particular network efficiencies. Sprint's network is in fact its biggest asset. The carrier does not have to worry about separate networks for data transmission since its voice network can handle the task. Sprint's biggest debit is its underdeveloped software for their network. Software development for its databases has been hindered by the internal priorities of fixing billing problems and launching a new billing platform. With the billing problem resolved, it is playing catch-up, especially in the features arena. While the carrier probably has one of the better networks, it has some way to go to match the service and feature options offered by MCI and AT&T.

- Telecom*USA has stressed its regional appeal, operating two distinct regional carriers – SouthernNet in the South and Teleconnect in the midwest. Each of these divisions has a substantial portion of fiber in their networks, which is attractive to MCI. Telecom*USA is known for its product development efforts, which have given rise in the past year to enhanced voice mail and travel card services.

Among the other providers are Allnet, ITT, Cable & Wireless, ATC/Microtel and Williams Telecommunications. Each of these carriers is pursuing their own niche markets. Williams Telecommunications is exclusively a private line provider and has established itself as a strong second-choice carrier behind AT&T for private line services. Allnet, ITT, and Cable & Wireless are all strong national communications providers with offices in major cities around the US; however, they clearly represent the second tier of long distance carriers as their network development lags behind those of the top three. ATC/Microtel is one of the most successful long distance carriers. The company operates primarily in Florida and in Texas and owns extensive fiber facilities in both of those areas. ATC/Microtel sells both switched and private line services.

The Dynamics of Today's Market

The telecommunications market is changing daily and is sometimes confusing. Yet, there a number of distinct trends in the industry that bear mentioning.

1. Emphasis on switched versus dedicated services. Ten years ago, it was a big deal for a user to have a vast worldwide private network comprised of innumerable dedicated circuits. Maintaining these networks involved a sizable staff and constant tinkering. Today, the emphasis increasingly is shifting away from global dedicated networks in favor of switched alternatives. The higher quality of the public switched network now makes switched alternatives possible for the first time. As the local and long distance providers continue to replace older analog and copper based transmission equipment with newer digital, fiber optic based facilities, the reliability of switched transmission increases multifold.

2. Integration of voice and data. In many firms, the age-old split between data communications and voice communications forces is blurring. As telecommunications services become more open to both voice and data over the same facilities, firms are effectively merging the functions together. The advent of ISDN promises standardization of this marriage of voice and data.

3. Integration of intrastate, interstate, and international products. Where multiple jurisdictional products existed in the past, seamless services are evolving for the future. State regulation is easing and the carriers are abandoning maintenance of separate product lines for intrastate, interstate and international services where possible, merging them into single, unified services under broad pricing plans. This makes network planning and design easier for the end user, and also gives each carrier a better chance at capturing all of a customer's traffic.

4. Creation of niche services. As competition becomes more intense, the carriers are planning vertical market oriented, niche specific services. Already, the carriers have started packaging specific product combinations for financial brokerage houses, real estate groups, and other such markets. This will continue in the future.

5. International extension of domestic services. As foreign networks become more sophisticated, US carriers are extending their domestic services abroad, in some cases by actually providing the foreign telephone company with the same proprietary software as used in the US. As a result, domestic firms with locations overseas can incorporate those sites into their existing networks.

6. Increasing mobility of services. Access to corporate networks is now greater than it ever has been. No longer must users be at a corporate site to take advantage of company long distance services. Through creative use of 800 numbers and calling card services, users on the road can call directly to corporate sites through hot-lines, upload or download facsimile messages that have been stored in customer-specific mailboxes, and electronically input orders for items to be shipped the next day. Cellular facsimile machines and data devices are now fairly common. Even the airlines are grappling with whether to make their cabin attendants become operators in the air, accepting inbound messages for customers in specific seat locations. Add to this advances in paging and other personal telephony technology, and users find it hard to get away from the office.

7. Continuing decrease in the underlying costs of carriers. Technological advances continue to allow carriers to cut their rates to end users. As network management and oversight becomes more automated, carriers are able to trim personnel and cut costs. Continuing migration to fiber and digital facilities also has contributed to cost decreases. Finally, the switch from largely physical connections to virtual software-based connections has reduced the effort involved in installing and altering customer networks.

8. Consolidation of the industry. While niche market firms continue to enter the long distance industry, the trend is still towards consolidation. The number of national firms is dwindling. MCI's purchase of Telecom*USA reduced the number of national competitors to under ten – including AT&T, MCI, Sprint, Cable & Wireless, ITT, Allnet, and the National Telecommunications Network (NTN).

9. Continuing decrease of high capacity circuit prices. As more fiber capacity is installed, the prices of T-1 and T-3 circuits continue to drop. These high capacity circuits are allowing more and more companies to carry bandwidth-intense applications, including videoteleconferencing and other image transfer programs. Fractional T-1 is available now and fractional T-3 is on its way.

10. Growing importance of value-added services. As the market moves towards maturity, enhanced value-added services, such as fax networks, 700 services, and travel cards, will provide the differentiating factors in the purchase decision.

Throughout this book, these trends will be discussed and related to present developments within each service.

Chapter 3

The Local and Long Distance Network

The public switched telephone network serves as the foundation for telecommunications today. It is the public facility through which most local and long distance calls pass.

While the legal and regulatory distinctions of "local" and "long distance" networks are fairly clear, the physical separations are not. Indeed, local calls can be physically completed through long distance carrier switches, and long distance calls may be carried entirely through the facilities of the local exchange carrier. While neither are legal in most areas of the country, physical interdependence between the local and long distance carrier networks is making it increasingly difficult to separate the service areas of the different carriers.

This is particularly evident the increasing reliance on computer-driven facilities. The installation of intelligent software programming to govern the latest central office based services is one of the causes of this interdependency. Newer, more advanced network signaling systems, including signaling system 7 (SS7), are being installed in local and long distance networks to carry intelligence across networks. As customer demands become more sophisticated, local networks have to "handshake" with long haul networks in a more seamless, efficient manner – further blurring the differences between previously distinct networks.

For instance, look at the routing of an 800 call today. Toll-free calls, originated by dialing a 800, followed by a seven-digit number, must go through several database checks before being routed by the network. One database query tells the network where to send the call. The call is not routed using the dialed numbers, but rather a translation of those numbers to the termination point's standard ten-digit public network telephone number. This is accomplished as the call enters the long distance carrier's telephone network. Typically, the originating switch screens the inbound call and sends a special signaling request to the main computer databases. Successive database queries could provide the network with supplemental handling instructions, such as alternative handling based on the time of day. With SS7, the signal requesting this information travels on a separate path than that of the call.

Not all carriers have their own translation facilities. For carriers without their own database capabilities, this translation could be provided by a local exchange carrier, another long distance carrier, or some other service provider. In these cases, the carrier must interconnect with another network for signaling exchange.

The end result is that, due to the sophistication of the networks, different parts of the call can now be going to different entities, over different paths. Thus, the concept of separate "local" and "long distance" networks is rapidly becoming obsolete.

So, just because a company buys service from one long distance carrier does not mean that carrier is providing that user with all portions of that service. For this reason, it is important to gain a solid grasp of the fundamentals of today's public switched telephone network.

The Local Exchange Central Office

Typically, there are three entities involved in the placement of a domestic long distance call – the originating local telephone company, the intermediate long distance carrier, and the terminating local company. Translating this routing into telecommunications terminology, these three portions represent the access, transport and egress of a call. There are many variations of this theme – for instance, in placing a long distance call to a town nearby, all three entities might be the same local carrier. Also, the originating or terminating entity might be a telephone company, a cellular carrier, a private satellite company, or any of a number of different entities. With international calls, one end of the call is handled by the foreign local exchange carrier, as is a portion of the interexchange circuit. For now we will stick with the simple model of a Local Exchange (LEC) to Interexchange (IXC) to LEC telephone call.

As mentioned in Chapter 2, telephone transmission in your local calling area is governed by your local exchange carrier. Each LEC has a series of interconnected telephone switches and telephone line trunk groups that accept and deliver telephone traffic. The arrangement of these trunks and switches is called a network configuration.

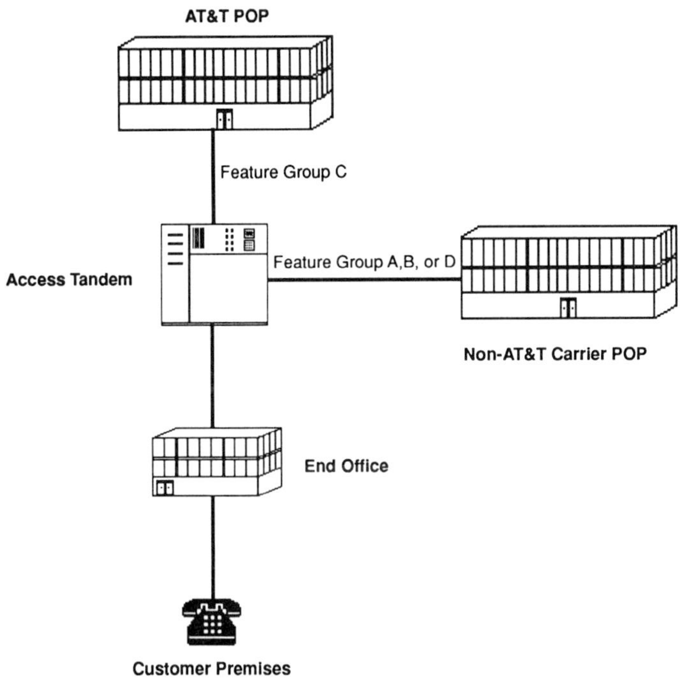

Figure 3.1 The equal access local network.

Configuring the public switched network is like designing a customer's private network. The local telephone companies optimize their network configurations on a continuous basis to keep costs to a minimum. Where possible, network architectures are designed to concentrate traffic over high-volume facilities for economies of scale, and network switches have their own optimization algorithms for least cost routing, that is the selection of the least expensive route at any particular time. Other factors, such as lines being busy and equipment failing, are also taken into account in designing these networks. Alternate routing paths are established so that the network knows how to react when a failure or busy condition arises.

Central Offices

The main component of the local telephone network is the central office (CO). You may also hear these referred to as "end offices" or "Class 5 offices." Central offices are typically the first point of entry into the local telephone network. Each office is characterized by a unique area code and exchange combination, such as 203-645 or 804-966. These are the first six digits of any domestic North American Numbering Plan (NANP) telephone number.

The capabilities of these switches define the overall functionality of the entire local network. It is here that your call is examined and routed according to the digits dialed whenever placing a call.

End offices have become a powerful stanchion of software-defined network services. Local telephone company advanced services, such as Centrex and basic virtual network screening functions (discussed later), are resident in these switches.

Remote Switching Units

Increasingly, remote switching units are supplementing end office switches, allowing the local telephone company to quickly expand service into growing areas without having to install full-size end office switches. The remote units can perform many of the same functions as the larger Class 5 switches, but on an independent basis.

These remote modules are critical for redundancy and disaster recovery in the local loop. Since they can provide basic local service, they can remain functional even when their parent central office is knocked out of service due to a flood, an automobile accident, etc. This alternate routing acts to reduce overall downtime of local service.

The Long Distance Carrier Point of Presence

The long distance carrier also has a key fundamental component in the local environment – the point of presence, or POP. This is the interexchange carrier's interface with local telephone company. Generally, each major carrier has at least one POP within each LATA where it intends to provide originating and terminating service. In major metropolitan areas, carriers may have more than one POP to cut down on many mileage-based access and egress costs and to improve accessibility to their customers.

Figure 3.2 A long distance carrier's point of presence (photo courtesy of US Sprint).

Points-of-presence can be a large as a national operations center or as small as a hotel room. A POP is simply a large computer frame with cross-hardware connect frames and software storage devices. Local exchange network termination lines bring traffic to, and accept traffic from, the carrier POP. Correspondingly, carrier long haul lines exit the building and carry traffic to and from the carrier's long distance network.

These POPs are governed by national operations centers that oversee the operation and maintenance of the equipment, mostly by remote monitoring equipment. Few carrier POPs are manned 24 hours a day; most are supported by off-site personnel that are dispatched when problems occur. This is another consequence of the increasing computerization of the telephone network.

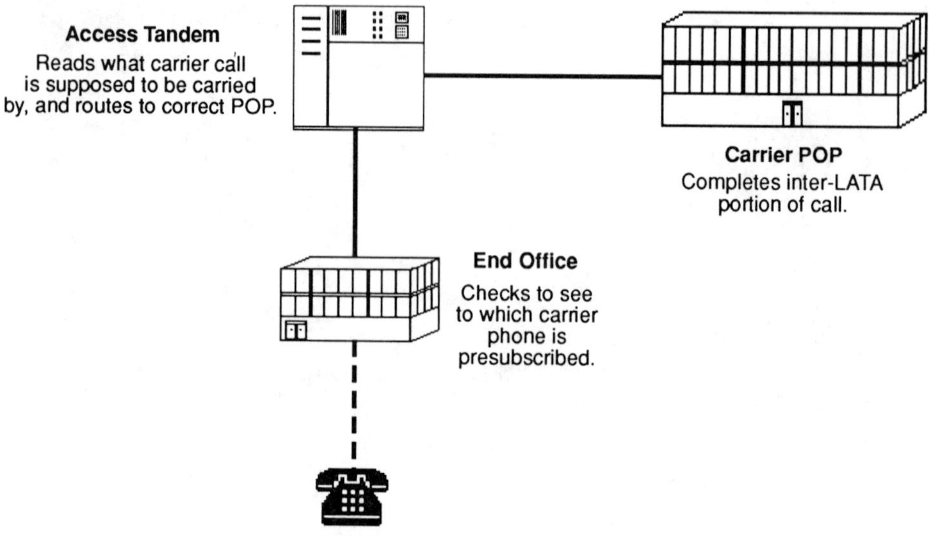

Figure 3.3 Switched access through the local network to a carrier point of presence.

The Long Distance Network

The long distance networks in existence today are truly remarkable when one considers the fact that most of them did not exist five years ago. Where AT&T had more than a century to develop its network, MCI, Sprint and others only had a few years. In that short period of time, tens of thousands of miles of fiber optic cabling has been installed nationwide, plus hundreds of carrier switches and supporting facilities.

While constructed in a relatively short time, these networks were not installed overnight – they evolved gradually. MCI's network, for instance, started with microwave facilities between just two cities (its original name was Microwave Communications Inc., hence MCI). One of the companies bought by MCI, Satellite Business Systems, used satellite as its primary means to transit voice and data communications. Few companies started off with advanced fiber facilities. Those who installed

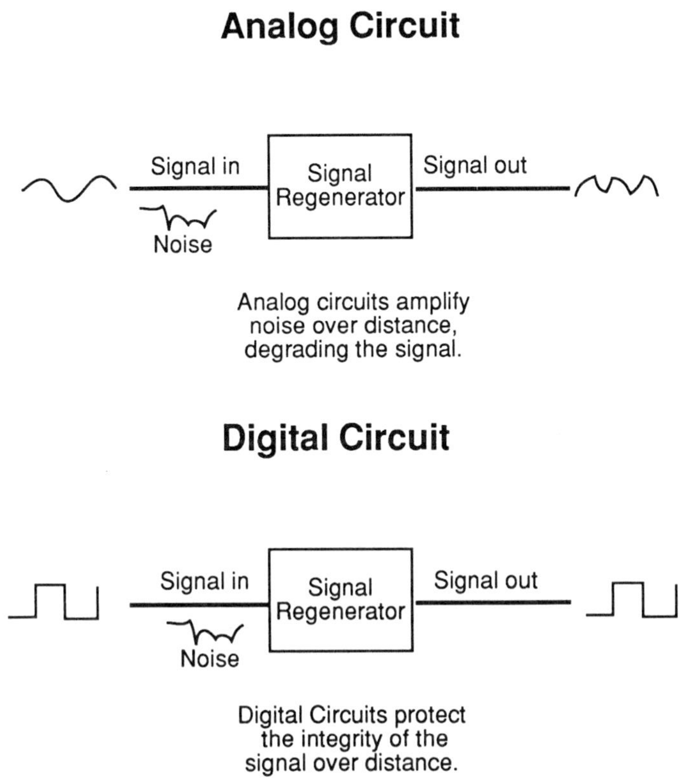

Figure 3.4 Analog versus digital signals.

fiber networks had the capital and long-term vision to begin the process several years ago. Only now, for instance, is Sprint seeing the operational and cost benefits of its fiber network program.

Today's networks are still a jumble of technologies. While fiber optic transmission is the preferred means of sending long haul messages, it is not always available or feasible. For instance, few companies see the economic incentive to route fiber cable to the top of Mount Washington, so microwave is likely used. Satellite, too, is popular, particularly in the international realm where transoceanic fiber facilities are subject to cable breaks. Copper cabling, for the most part, is not used in the long haul networks anymore.

The networks have more advanced transmission encoding technology too. Digital transmission is the preferred mode today. Analog facilities are being phased out where possible. The advantages of digital versus analog are many but the major impact is in signal regeneration. Most long haul signals must be regenerated at some point in the transmission between the originating and terminating points. With the older analog technology, not only were the inbound signals regenerated, but also any accompanying noise (static, crosstalk, etc.) that is detected. Digital systems only regenerate a 0 or a 1 – any accompanying noise is dismissed. Thus, the signal reaching the endpoint is nearly as clear as at the originating station.

The emphasis on digital technology is also spreading into the local loop where supporting systems are being established. It is far easier for the interexchange carriers to implement digital technologies in their hundreds of POPs and interconnecting facilities than for the local carriers to install it in their tens of thousands of end offices.

The Local Loop – Accessing the Long Distance Network

Competition in the local exchange environment is spawning a wide range of access and egress options for an end user with a long distance network.

As mentioned, the most common means of access is through the local exchange carrier. This means routing a call through the local telephone company's network for delivery to the carrier's POP. LEC access is divided into switched *versus* dedicated options.

Switched Access

Switched access simply means the call is switched via the public network to the carrier's POP. Each time a call is placed, the route the call travels is determined by the facilities available in the public network at that particular time. Thus, in one instance, a call might route directly through two end offices to reach a POP; a minute later, the same call could pass through four end offices due to busy facilities.

At present, there are four types of switched access – called feature groups A, B, C, and D. The type of connection available to the end user in any particular geographical area depends on the type of access purchased by the carrier. However, there are vast differences in line and performance quality among the different options so it is important to realize what each feature group offers.

Feature Group A: As the least expensive and also least sophisticated form of access, feature group A was typically the first mode of access offered in new marketing areas as AT&T's competitors spread nationwide in the mid-1980s. Feature group A does not offer the 1+ dialing discussed in Chapter 2. A customer must dial a particular local telephone number that connects to the local telephone company. The sub-

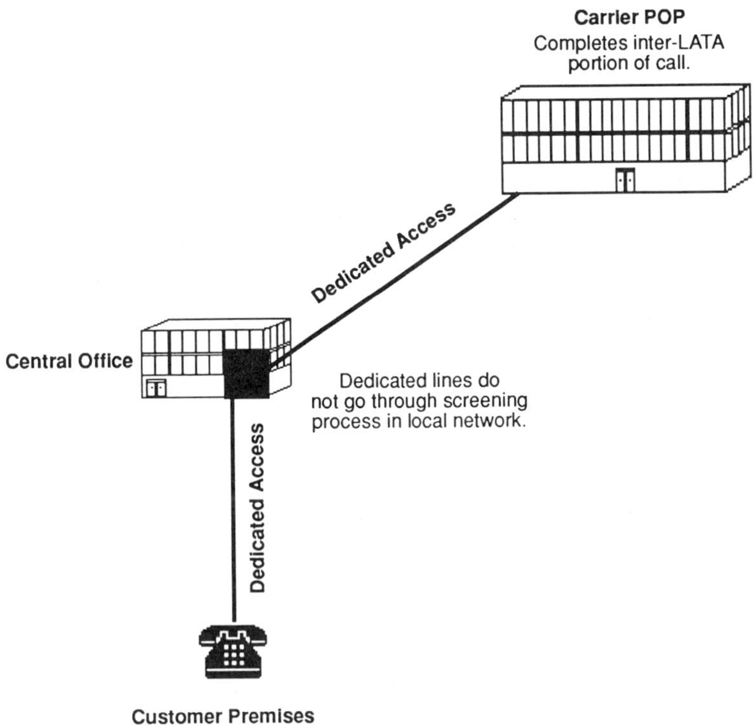

Figure 3.5 Using dedicated access through the local network to a carrier point of presence.

scriber then hears a tone or announcement from the carrier asking the user to input additional user identification and telephone number digits. In all, the process can require 23 digits to complete a ten-digit phone call.

The user is required to enter these extra digits because feature group A circuits do not provide automatic number identification (ANI), that is the telephone number of the calling party, to the carrier

Business customers might shortcut the process with autodialers, equipment which pulses the required identification digits after a user dials a telephone number. Historically, however, autodialers have been largely unreliable.

Feature group A access to the local network by the carrier is through a two-wire, "line-side" connection. Two-wire lines are the standard configuration for residential telephone service. Line-side denotes quality equivalent to the local network provided to the standard, two-wire subscriber local lines. Higher "toll" quality lines have four wires.

A drawback to feature group A connections is that local "message unit" charges may apply. Message units are fees assessed by the local telephone company for placing local phone calls. Whenever a user pays 10 or 20 cents at a payphone, a large portion of this goes towards message units for placing a local call.

Message units may not be used in your area. Many local exchange companies offer unlimited calling for a flat price per month. Others, such as Southern New England Telephone, have multiple plans. In these cases, low volume users might want to pay on a per call basis. Higher volume users, particularly businesses, can purchase unlimited calling options in their local area, so they will not be charged for any calls to a long distance provider's feature group A local number.

However, other local jurisdictions do not have such unlimited options and users pay strictly on a per-call basis. In Connecticut, that means that 12.5 cents is added to the cost of each call.

Feature Group B: While feature group A circuits are not so common anymore, feature group B circuits are. A user gains feature group B access by dialing 950-0XXX or 950-1XXX, where XXX is the carrier identification code allocated to the carrier by Bell Communications Research (Bellcore). MCI customers, for instance, obtain access by dialing 950-1022. Once the MCI network is accessed, the user gets a tone or announcement for extra digits, just as with feature group A.

The advantage of feature group B circuits is better quality. Group B lines are four-wire, trunk-side connections – meaning that nearly double the bandwidth capacity is used to carry the same calls as under feature group A. Also, 950-calls are toll-free in the local area. Message units are not a concern to users. The only exceptions are when a user places calls from hotel rooms, private payphones, or any other situation in which a user may be charged for using the telephone.

Automatic number identification is available with some forms of feature group B access. However, since the long distance carriers cannot always rely on the availability of this ANI signaling, they have chosen to require user identification codes all the time, just like feature group A circuits.

Feature Group C: AT&T is the only carrier that can use feature group C access and this is a holdover of pre-divestiture days. Feature group C access offers 1 plus dialing and four-wire trunk-side connections. Automatic number identification is available from nearly all feature group C circuits.

Importantly, the feature group C lines run directly from the end office to AT&T's POP, giving the carrier's users a high degree of line quality and extremely short call establishment time. AT&T's competitors, using supposedly equivalent feature group

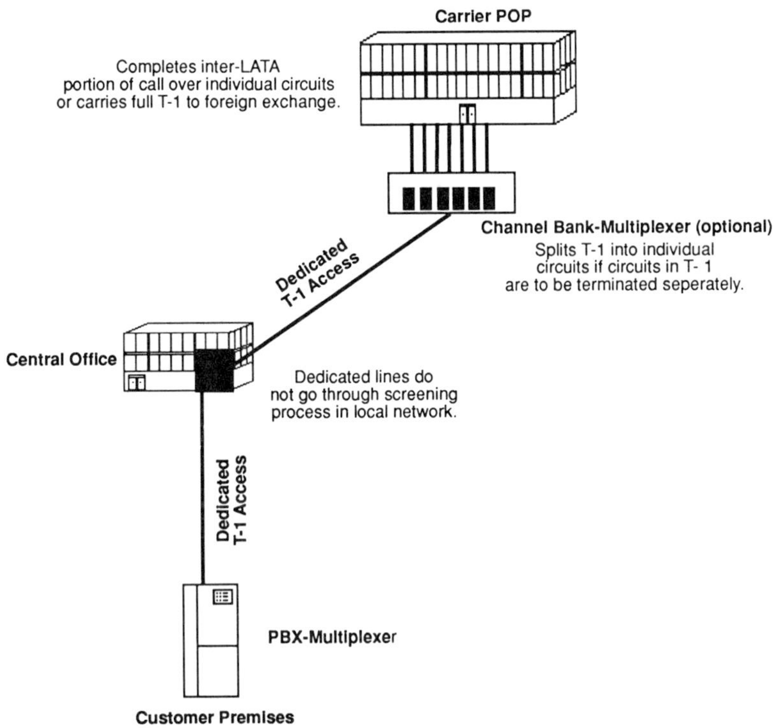

Figure 3.6 T-1 access allows access to a number of carrier services over a single circuit.

D circuits, suffer from slightly longer call establishment times because these calls are routed through the public network to the carrier POP – there typically are no direct connections.

Feature group D: Feature group D represents the equal access circuits. This form of access is the foundation of modern long distance competition. Feature group D circuits are four-wire, trunk side connections that are supposed to offer quality comparable to feature group C. It features 1 plus dialing and ANI in most locations.

These four forms of switched access are typically used by the smaller to moderate-sized customers. This size of customer generally cannot justify the higher line costs associated with dedicated access, even though dedicated access long distance products generally have lower usage costs.

Dedicated Access

The other major form of access in the local loop is dedicated access. This means that a user has purchased a line specifically for its own use between the customer premises and the carrier POP. This path is reserved for just that one user – the path of a call one day matches the path of a call the next day.

Dedicated access is typically reserved for higher volume customers. When customers subscribe to switched access products, the carrier incurs per minute usage costs for each originating call. Naturally, these costs are passed onto the customer in the form of higher usage rates. With dedicated access line (DAL) facilities, there are no per minute access costs – these are covered by the monthly cost of the line itself, and so usage costs can be lowered. Since dedicated access facilities can be expensive, sometimes more than $120 per line, usually only the larger businesses sign up for the services.

There are two general classes of dedicated access connections, also sometimes referred to as "special access" connections. The first is the WATS access line (WAL). This is the traditional method of access that has been used for years to access AT&T's standard WATS product. A WAL is a dedicated connection from the customer premises to the local exchange carrier's central office. However, from the central office, the call is routed through the public switched telephone network like any other switched call. This necessitates ending the dedicated connection into an equal access central office – without that capability, the end office would not be able to determine to which carrier to send the call. In cases where the central office is not equipped with equal access, a WATS access line extender (WALE) extends the dedicated connection to the nearest equal access office.

The second form of dedicated access is the dedicated access line (DAL). This is the standard form of connection for most large companies. As with the WATS access line, the connection between the customer premises and the serving central office is dedicated; however, with a DAL, the portion from the serving central office to the POP is likewise dedicated – giving the customer a reserved channel the entire distance.

In most cases, the dedicated access line is desirable for carriers and users alike. Since the switched network is a shared network, no one can predict the exact path a call will take each time it is placed. With dedicated facilities, the path is predetermined – making it easy to monitor and maintain. For users with high speed data applications, dedicated connections are the best way to go.

However, there are instances when WATS access lines are preferable. Certain states, such as Texas, have more favorable WAL rates than DAL rates, simply because that's the way the present tariffs are filed. In these instances, it is more economical to install the WALs.

One drawback to DALs can be cost. Depending on the distance between the carrier's switch and the serving central office, the actual cost of a dedicated line could be high. Luckily, most carriers will guarantee the cost of a line as long as it falls within the local distribution area of the carrier's switch, for example within ten miles.

T-1 Dedicated Access Connections

Changing economics in the local networks are making certain forms of DAL access more appropriate for medium- to high-end users. Local telephone companies offer high capacity digital access lines, called T-1 and T-3 access lines, that are 24 and 672 DALs respectively. T-1 access is becoming quite popular, especially given the plummeting costs of IXC fractional T-1 and T-1 private lines, discussed in chapter 7.

Although T-1 access lines might seem expensive, several hundred dollars in most areas, they are not hard to justify relative to dedicated access lines. Typically, the cutover point between DALs and T-1s can be as low as 4-6 DALs. In such instances, the user at the cutover point is effectively getting 18-20 DAL lines "free" by switching to a T-1.

The 24 individual channels within a T-1 access line may be divided among several different services – outbound WATS, inbound 800, international services, fax networks, private lines, etc. Furthermore, today's sophisticated network management tools allow users to dynamically allocate these channels among the different services from a terminal at the customer site. Thus, if a user is not using all of the channels allocated to WATS traffic and needs an extra one to handle an 800 promotion, the user could simply re-allocate a WATS channel to 800.

Intra-LATA Screening

Recall that the lines of demarcation between the local and long distance networks are supposed to be absolute. These definitions again run into problems when talking about dedicated connections from the customer premises to the carrier POP. Since dedicated connections run directly between the customer and the carrier POP, the local carrier has no chance to screen calls to see if they are destined for intra-LATA points.

Most states have ruled that this intra-LATA long distance traffic must be carried only by the local telephone company. However, many users do not have the means to separate the traffic between the local and long distance carrier. Thus. for directly connected users, this responsibility falls to the long distance carriers.

With equal access switched (feature group C or D) traffic, there is no problem. When the traffic passes through the serving central office, the local telephone company screens the first six digits of the call to determine if the call is destined for a point outside of the LATA. If so, the call is sent to the carrier POP. If not, the call is captured by the switch, routed through the local exchange carrier's network, and billed by the local telephone company with the monthly bill.

With traffic originating over dedicated facilities, the process is much more difficult. The path between the customer premises and carrier POP is basically a private line. Since the local telephone company never actually "sees" the call as it passes through its switches, there is never a chance to screen the call for intra-LATA termination. That is, the carrier could receive the call in its switch over dedicated facilities and then send the call back into the local network without telling the local carrier.

State Public Utility Commissions have dealt with this in two generic ways, commonly referred to as "block or pay" rules. The IXC has the option of blocking the call, and handing the call over to the local carrier, or carrying the call over its own facilities and reimbursing the local exchange carrier for its lost intra-LATA revenues.

Most long distance carriers do not want to have any calls blocked on their network. It just does not look good to customers. So, for the most part, the calls are carried by their switches and the local exchange company is reimbursed for the lost toll revenues.

Other Forms of Access

Generically, feature group D and dedicated access lines comprise the majority of access into the long distance carrier point of presence. However, there are a few other notable ways in which customers can reach an IXC network. Some of these combine the feature group and dedicated access options with other known technologies. Others are stand-alone access options. Not all are supplied by the local exchange carrier.

Centrex

It is time for all managers to get familiar with Centrex. Centrex is being billed as one of the primary access options of the future, if not today. The BOCs are placing a great deal of resources behind central office based services with the expectation that this will give them a firm foundation against future competition by the interexchange carriers. This makes the Centrex access line a link to the local exchange carrier's central intelligence hub.

In concept, Centrex services provide the user with the same functionality as he or she would get from a private branch exchange (PBX) – call forwarding, call waiting, call hold, etc. In this manner, Centrex competes with the PBX equipment manufacturers.

Under PBX systems, telephone sets are connected to a private switch on the customer premises, which in turn is connected to the serving central office via standard business lines. Under Centrex, each telephone on a user premises is connected directly with the Centrex CO.

Centrex services fall into one of two general categories: as stand-alone services or as enhancements to local service. The distinction comes in whether a user's regular business lines must be converted physically to Centrex lines, or if Centrex functionality may be merely added to the business lines.

How does this affect long distance? Typically, an end user will designate two or more levels of access for long distance with Centrex. A caller would dial 9 to make a local call and 8 to make a long distance call. This is similar to the ways that most PBXs are set up.

Centrex customers are able to establish varying levels of screening on each of the Centrex access lines, creating calling restrictions for each user. Customers can also program the Centrex to perform differential routing depending on the time of day or the day of week, etc. This can be determined by the routing tables within the system. The effective result is that Centrex services can act as a form of intelligent access manager to long distance carrier services. Centrex also adds considerable functionality to telephone service. Voice mail is now practically a standard offering for Centrex.

ISDN

Perhaps more important, however, is the role that Centrex will play with integrated services digital network (ISDN). In the early offerings, ISDN has been tied to Centrex services by most of the BOCs. This is not to say that ISDN is only available through Centrex; for example, AT&T has ISDN primary rate interface available in 323 cities nationwide.

ISDN is a critical form of access for long distance service because it opens a whole other level of network intelligence to end user. Among the services that are available with ISDN:

- **Switched 64 kbps** – This is the ability to send switched data transmission at 64 Kbps speeds through the public network. This is essential for international businesses that find problems in merging the US standard of 56 kbps with the global standard of 64 kbps.

- **Call-by-call selection** – This is the ability to select the long distance service desired with each phone call made. Typically, the present access options require the customer to designate in advance a service for each access channel. Thus, one channel might be for WATS, one for 800, etc. With ISDN lines, however, a user could conceivably have access to a wide range of voice, data, image, and facsimile services. Each call would contain off-line signaling that would tell the network what type of call was coming and which service to use. The savings in access would be enormous.

- **Automatic number identification** – Again, through the use of out-of-band signaling, the ISDN line can provide valuable information about each call, as it comes in. One such piece of information is Automatic Number Identification (ANI). This tells the called party the telephone number that originated the call.

- **Voice and data integration** – The multi-channel ISDN lines can carry voice over one part and data over another, allowing end users with PCs to talk and send data back and forth simultaneously.

ISDN will establish an international standard that promises to synchronize the world's telephone networks. By standardizing interfaces, information will flow more directly, bypassing many of the translations that must take place today. Because ISDN is digital, it will provide a higher degree of accuracy in transmission.

ISDN access lines are provided by the local exchange companies. There are two ISDN access interface formats: the Basic Rate Interface (BRI) and the Primary Rate Interface (PRI). With BRI, there are two 64-kbps bearer channels, called B-channels, and one 16-kbps data channel, called the D-channel. This is commonly called "2B + D." PRI access provides 23 bearer channels and one data channel for 1.544 Mbps of transmission; a European equivalent has 30 bearer channels and a single data channel. These are called "23B + D" and "30B + D" respectively.

The bearer channels are designed for multiplexed voice, data, group 4 facsimile, videoteleconferencing, or whatever type of digital stream offered to the circuit. The data channel carries information about the calls. It is also used for access to slow-speed data networks, including packet switch and videotex networks.

Other Means of Local Access

Buying service from the local exchange carrier is not the only means by which a company can acceess their long distance carrier. There is a wide range of other options, including cellular, bypass providers, or private networks. These are discussed in detail in the next chapter.

Chapter 4

Alternatives to the Local Network

Up to this point, it has been assumed that the local telephone company would be involved, at least in some manner, in providing access to the long distance carrier point of presence. This is not always the case. The monopoly of the exchange carrier over all local services is diminishing. Alternative forms of access are increasing and end users now have viable alternatives for getting their calls to the long distance carrier POP.

One cause of this surge in alternative access has been the continually high price of access from the local exchange carriers. One of the primary reasons given by customers wishing to bypass the local exchange carrier has been price. Other factors listed as important include flexibility, responsiveness and control.

Bypass Options

Perhaps the most common form of bypass is replacing the premises to POP link with private facilities owned by the customer. Advancements in various transmission technologies have lowered the cost of private access, making microwave, infrared, fiber optic, satellite and copper cable bypass quite cost-effective.

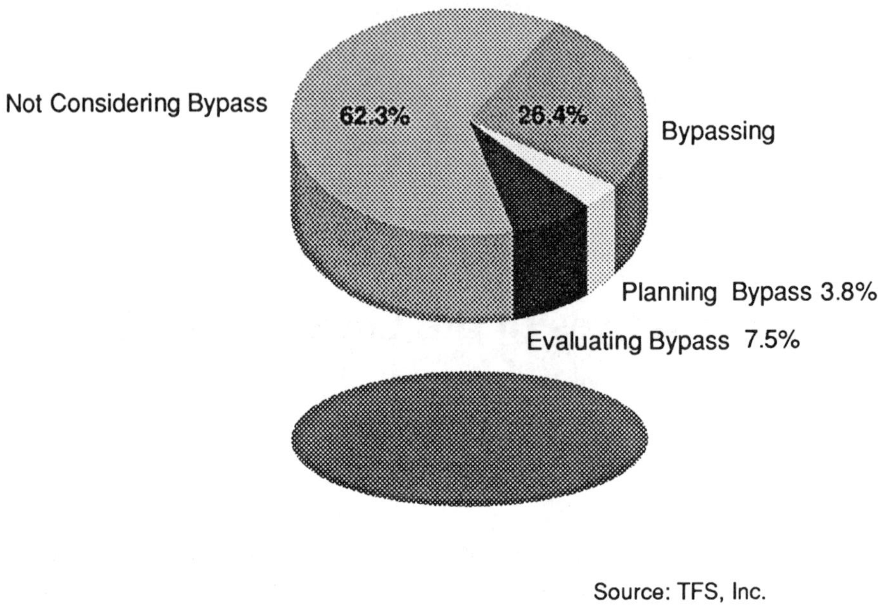

Figure 4.1 Survey of companies considering or using bypass.

Microwave

Microwave links are perhaps the most common form of bypass today. Microwave radios are designed to transmit voice, data, facsimile, image and video between two points, over a variety of frequencies. The most popular models of microwave equipment transmit in the 6- and 8-GHz band. However, frequency allocation in these frequencies is getting overcrowded, particularly in major metropolitan areas where bypass is frequent. This has forced users to examine other frequencies. Lower frequency microwave systems can transmit messages over fairly long distances, but in general, require larger antennas and higher up-front investment and installation costs. Higher frequencies, on the other hand, can only transmit over shorter dis-

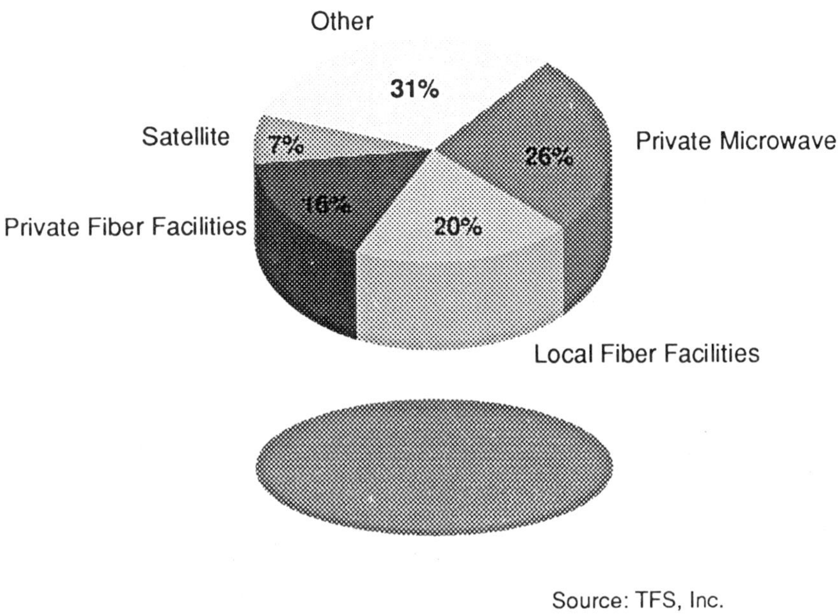

Figure 4.2 Bypass Options.

tances, but cost less and have smaller antennas. Also, available frequency space is more abundant.

For these reasons, many bypassing companies have traded distance for cost and convenience by opting for the high-frequency 18- and 23-GHz equipment. After all, most access lines are less than five miles in length – well within the capability of this equipment. In fact, on good days, 18- and 23-GHz microwave transmitters can send signals up to 12 miles away. Poor weather, however, can drop this distance to just 4-5 miles.

A clear line of sight is required for microwave links. Signals will not pass through buildings or go around corners. Additionally, the line of sight needs to be at least 30 feet in diameter to ensure unobstructed transmission. Signals too close to buildings or other hazards can suffer from reflection, which corrupts the overall signal.

Figure 4.3 Microwave systems are a familiar sight in most metropolitan areas.

In instances where customers have to pass signals through buildings or over longer distances, multiple stations might be needed. Repeater points would receive and retransmit signals, allowing the transmission to "hop" to the termination point. In practice, too many hops make the project too expensive to contemplate.

For many users, microwave systems may be less than ideal. Among the frequently cited problems are:

- **Installation.** Microwave systems are not too portable, especially given the extensive engineering that must be performed for installation. The FCC must license the installation of all microwave systems to protect all other microwave users in the area – competing systems can quickly make a microwave network useless.

Licensing is not as painful as it could be. An engineer from a frequency coordination group must complete a spectrum path study that shows the new system will not interfere with others already in operation. FCC approval is typically swift after these studies are submitted. In all, the entire process generally takes about three months – the same amount of time its takes to get most local T-1 access lines installed.

- **Zoning.** Some local zoning boards have passed superbuilding ordinances that limit the placement of microwave equipment on customer buildings.

- **Environment.** Microwave signals in the 18-GHz and 23-GHz frequency bands can be weakened by rain, a phenomenon called attenuation. In general, problems from rain attenuation are often overstated, mostly by competing technology vendors. Still, it is a problem for longer distance transmissions. On days which there is a moderate amount of precipitation, communications signals will start degenerating after about five miles. In a heavy downpour, this figure can drop to 2-3 miles.

Most of these problems are taken into consideration during the initial design phases and, after installation, do not prove to be a problem for most companies.

Infrared

A strong competitor to microwave in the 1-3 mile transmission range is infrared. Infrared is generally less expensive than microwave and it is also more portable.

Infrared systems consist of a light source, such as an light-emitting diode (LED) or a laser, that operates in the infrared spectrum. Inbound signals are received by a photodiode which translates the signal for serving communications devices.

As with microwave, a line of sight is required, although infrared does not require the wide coverage path of microwave. The light beam is quite tight over the distance of the signal – the dispersion is less than five feet over the length of a mile. This characteristic makes infrared ideal in areas where congestion in common. In fact, multiple infrared systems can operate in the same area – something microwave systems cannot generally do.

Infrared's best applications are those which require transmission less than .5 mile; however, infrared transmissions can reliably travel up to three miles in length. As with microwave, infrared can transmit from rooftops or windows but is generally unaffected by superstructure zoning regulations because of the small equipment size.

Above all, infrared systems are easy to install and inexpensive. A low-end system costs as little as $3,000; T-1 capacity systems cost around $10,000 and can be installed in a few days.

Infrared suffers from some of the same environmental problems that plague microwave. After all, infrared is a light beam and anything that affects the travel of light will disrupt infrared transmissions. This includes fog, smoke, air pollution, clouds, etc. Heat shimmer due to asphalt will also distort the transmission – something to think about in metropolitan areas. Finally, paths that share a direct view of the sun rising or setting or that intercept reflections off of buildings may be unsuitable for infrared as the light overwhelms the photodiode.

As with any well-engineered system, however, these problems may be minimized with proper attention during the design phases of the program.

Satellite

Satellite has been perceived as a rather extravagant form of bypass for years. Companies use satellite dishes to access private network circuits provided via Comsat, Intelsat Business Services (IBS) or a broad range of satellite communications services from major satellite teleports.

These users bounce signals off satellite transponders for multi-site data transmission and voice broadcasts. However, due to the delays incurred by "hopping" across satellites, voice transmission may or may not be practical for particular user applications.

There are many advantages to using satellite as a means to bypass the local exchange carrier network. They are ideal for point to multipoint applications, such as data broadcast. It is also perfect for accessing areas inaccessible by standard infrared, microwave, or cable means – maritime or mobile locations, for instance.

Satellite used to be reserved for the largest companies because it was so expensive to build or rent the appropriate satellite facilities. However, technological advances in the very small aperture terminals (VSAT) have brought satellite transmission prices down and increased the applications potential of satellite bypass.

The small 4-foot VSAT dishes are now a regular feature of corporate rooftops across America. A major advantage of VSATs is control. VSAT users are not dependent on interexchange carriers for the long haul transmission – outside the space segment, the user retains complete control of the network.

This control generates a degree of flexibility since new sites can be added and old sites deleted within a relatively short period. Adding sites merely involves the installation of another site dish and its associated equipment.

VSAT networks are available in one-way or two-way configurations. By their nature, one-way systems are limited to data and video communications from a central site to remote sites. Many auto dealerships are using VSAT for maintenance and management training sessions, as well as data distribution about parts availability and new car shipments.

Two-way systems allow for interactive communication. This is a relatively recent technological phenomenon and the one that is likely to spark the greatest growth in the marketplace. Due to the size of the VSAT dish, remote VSATs cannot communicate with one another – they must use the central site as a relay for sending information from one remote site to another. The biggest applications for two-way VSAT are point-of-sale transaction data delivery and credit verification transmissions.

VSAT's operate on two frequency bands assigned by the FCC – the C-band ranges roughly from 4-6 GHz and the Ku-band covers 12-14 GHz.

The C-band is less affected by inclement weather than the Ku-band. However, C-band networks are susceptible to interference from other satellites and from terrestrial microwave transmission systems. Thus, they require some of the same licensing procedures found with microwave.

Ku-band systems are more sensitive to weather conditions. However, their higher frequency allows the use of smaller antennas and the relatively vacant state of the Ku band makes licensing considerably easier than C-band systems. This also makes the lead time for installation of Ku-band VSAT systems considerably shorter.

The physical installation of a VSAT system only takes about one-half of a day and is generally done by the vendor.

Fiber and Copper Cabling

Certainly another private bypass option is for users to install their own fiber or copper cabling between the premises and the carrier POP. Definitely fiber optic technology is the preferred technology when budget, time and rights of way are of no consequence.

This is unrealistic, however – these are standard considerations in any network and managers should realize that fiber optic systems are expensive and require legal rights of way through the property which the cable may run. This property is expensive and time-intensive to arrange. Extensive fiber and copper networks are thus limited to the very largest firms.

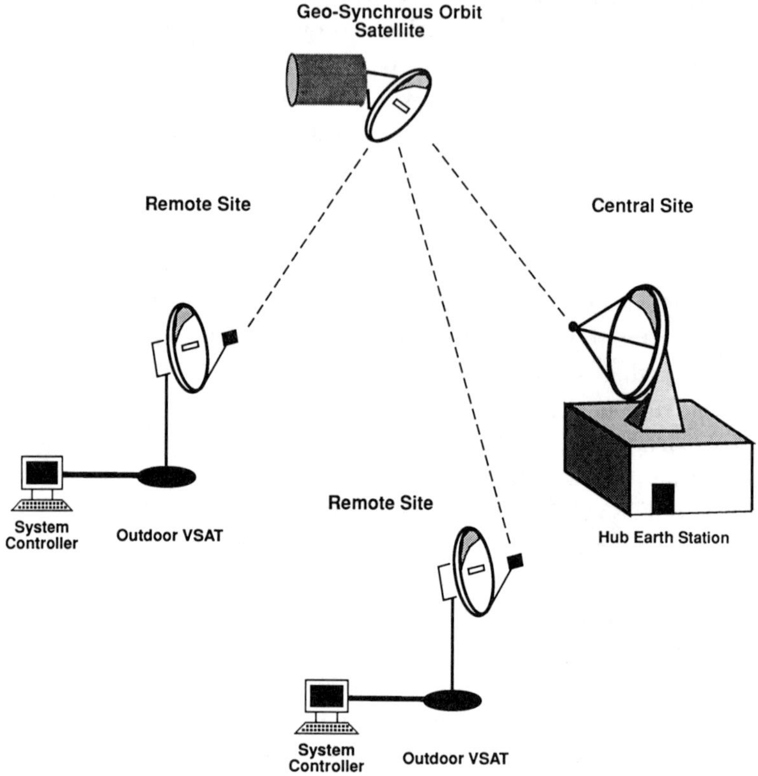

Figure 4.4 VSAT networks are relatively simple to understand and install.

Fiber optics is likely the choice of most managers in terms of transmission media. Fiber lets users transmit: signals further than microwave or infrared without a repeater; at speeds higher than 1 Gbps; free of environmental concerns, such as fog, rain, and snow; and, practically free from error. Fiber's major problem is installation costs. The cost of a private fiber bypass system in a non-metropolitan area can start at $100,000 per mile, with metropolitan costs hitting $500,000 per mile and higher.

Much the same can be said of copper cabling since the rights-of-way and construction considerations are very similar. Twisted copper pairs are only used for the shortest distances; fiber is generally more cost effective for the longer hauls. Fiber is also more expandable for future applications. It is no mistake that the local and long distance carriers are laying mostly fiber these days and not copper cable. Nevertheless fiber cable is not generally an option for most firms.

Metropolitan Area Networks

For companies that are lucky enough to be located in an area served by a metropolitan area network (MAN), the private network decision becomes slightly more complex. A MAN will typically offer the user a level of quality and reliability equivalent with that of a private network, since most are fiber-based and their routes are designed to provide high levels of diversity.

Commercial metropolitan area networks are presently installed or under construction in Washington, D.C, New York City, San Francisco, Chicago, Milwaukee, Orlando, Indianapolis, Philadelphia, Boston, Tampa - St. Petersburg, Detroit and Houston. In some cities, there is more than one metropolitan area network in operation, giving users the added benefit of price competition.

Metropolitan area networks have a great deal to offer the commercial user. MANs provide much of the same functionality as a private fiber network but without many of the hassles. For instance, with MANs, customers:

- **Do not have to acquire rights-of-way.** Buying property and arranging networks around available fiber routes is expensive and full of headaches.

- **Can share the MAN's economies of scale.** Fiber costs the same per foot whether one or one hundred customers use the fiber. With a MAN, users get the advantages of economies of scale so circuit prices are less expensive. MAN retail rates are generally 10-20 percent below the prices of equivalent local exchange carrier facilities.

- **Can turn up circuits faster.** A T-1 circuit can be installed in as little as 48 hours with a metropolitan area network provider, but a private fiber network can take years to install.

- **Can gain flexibility.** MANs are typically small companies so they can be more flexible in meeting user needs. If a location is not on the network, MANs will generally be quite flexible in constructing fiber spurs if the fiber can be used for others within the building as well.

- **Can access key network information.** Many MANs will give the users direct access to alarm and telemetry information so that customers can monitor the network as needed.

The decision often comes down to 1) whether a metropolitan area network exists in the area, 2) does the MAN cover the necessary routes, 3) is the level of quality high enough for one's network and 4) can the user part with control over the physical circuits themselves. In general, if a user is located in one of the cities in which a metropolitan area network is located, chances are that the MAN will be less expensive than building one's own private network.

The Future of Alternative Access

The LECs have not failed to notice the expansion of competition in the local marketplace. In fact, in many jurisdictions, it has spurred a competitive response – in price and in network construction.

Look at Illinois Bell for instance. One of the first metropolitan area networks was constructed in Chicago. Illinois Bell responded starting in 1983 with fiber installation around the city. By the end of 1989, more than 90 percent of Chicago's metropolitan central offices were linked by fiber.

The opening of more intrastate jurisdictions to competition, will force the local exchange carriers to be more responsive to user concerns. For users, the result of this competitive atmosphere will be both lower prices and better networking options.

Chapter 5

MTS and WATS Services

For as long as there has been telephone service, basic long distance, commonly known in the telecommunications industry as Message Telecommunications Service (MTS), has been an integral part of most users' long distance networks.

It is this basic service with which most callers are familiar as residential customers. A user simply picks up the telephone, dials a ten-digit or international long distance number, and the cost of the call is included with the monthly local telephone bill.

For corporate users, MTS has been a vital part of company telephone networks. This has been particularly true for smaller businesses and branches of larger corporations which did not have the volume of long distance traffic to justify traditional dedicated WATS services. Also, major corporate networks have used standard MTS services as overflow when their high-volume WATS and private line network services were busy.

Over the past several years, however, the situation has changed dramatically. In a sense, the reign of the MTS service has come to an end. One could say that MTS as a basic long distance service is extinct.

In their place, WATS services have come to be the dominant service for small to medium sized companies. Even for the smallest end users, MTS is no longer an economical option.

MTS Ages Poorly

Competition is the source of these changes. Until a couple years ago, WATS services were not an option until customer usage reached at least $200 dollars a month. Smaller customers were left with no option except for MTS for their long distance calling. Volume discounts were something reserved for companies with "corporate" calling.

Today the carriers have largely abandoned MTS services in the face of an increasingly competitive market and more demanding customer base. The MTS services are barely appropriate for the smallest amounts of long distance usage. US Sprint, for instance, lists its basic Dial 1 product as applicable for the 0-1 hour per month target market of customers!

Carriers have always targeted MTS services towards the low volume residential market. One frequently hears of the 80-20 rule in long distance – 80 percent of the revenues are produced by the top 20 percent of the market. Conversely, the remaining 80 percent of the market – including all of the residential and small business traffic – produces only 20 percent of the revenues.

Competition is forcing the carriers to offer their WATS services to the lower end of the market. Users want lower costs and the WATS services are the easiest mechanisms available. As a result, the remaining MTS services are being priced to extinction. The most notable feature of the pricing is the full-minute rounding. In the first days of competition, calls were rated by two rate elements – a rate for the first minute plus an different, usually lower rate for each additional minute thereafter. Carriers put a higher price on the first minute to recoup some of the costs involved in preparing a call path through the network, the call set-up process.

Improvements in technology, especially the installation of signaling system 7, and competition have whittled away at this rating format. Now, all domestic calls are generally rated according to an initial period of either 30 or 18 seconds, plus an additional period rate measured in six second increments. Most of the time, there is no difference between this initial period and applicable multiples of the additional rate period – that is, the initial 18-second period is equal to three times the additional 6-second period. This means that the carriers no longer charge for call set-up.

The differences in net costs between a full minute rounding scheme and that of a 30-second/6-second scheme can be significant, as shown in Table 5.1.

MTS rates have the highest rates for long distance service. Per minute rates average around 25 cents per minute.

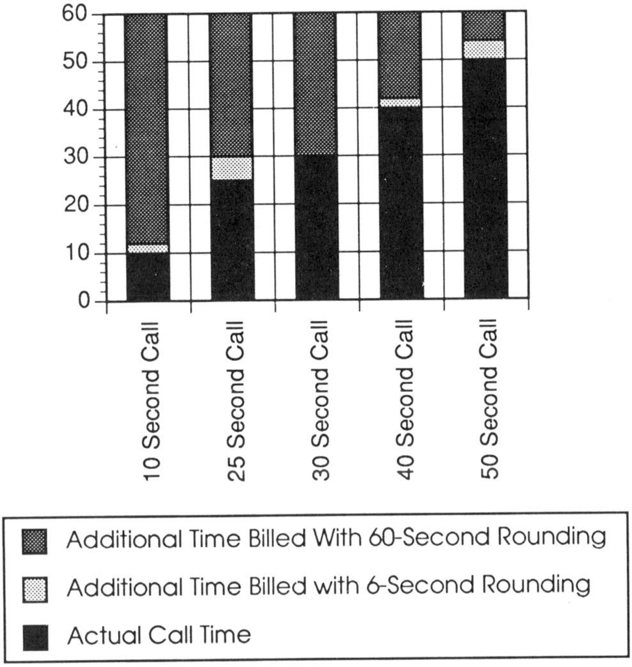

Actual Time of Calls:	155 Seconds
6-Second Rounding Billed Time:	167 Seconds
60-Second Rounding Billed Time:	300 Seconds

Figure 5.1 Call rounding can play a significant role in the cost of a call.

Thus, MTS services are not widely used. Casual callers (i.e., those using 10XXX equal access) and extremely low-volume users are the only remaining customers that realistically should be using MTS.

Stepping Up From MTS

All other low-volume users are better off if they sign with an optional calling plan of some sort. These programs are targeted towards specific niche markets and represent attempts by the carriers to segment the low-end of the market into competitive strongholds and to create tangible differences between otherwise generic products.

AT&T started these optional calling plans with its Reach Out America program. Since the inception of the Reach Out America program, all of the carriers have responded with similar programs that basically provide discounts for select types of usage, say calling after eleven o'clock or to a specific area code. Optional plans have become popular in the international arena as well, where the carriers have launched country- or region-specific discount plans.

Still, it is in the low-end that long distance has become truly generic. Profits are extremely slim and the extent to which optional plans can cut long distance costs is fairly limited. Increasingly, carriers are turning away from competing on price in favor of offering enhanced value-added services for end users.

700 Services

One of the novel ways that carriers are seeking to differentiate their low-end service is with 700 service.

Like 800 and 900 services, 700 service is distinguished by the prefix dialed by caller to reach the service. When an equal access call passes through the local telephone network, it is usually screened by the first switch in the call path. If the switch reads a "700" as the first three digits, this call is automatically routed to the carrier to which the customer has presubscribed the local line. This 700 call is then screened by the carrier switch, its POP, which then routes the call to the appropriate 700 service facilities.

These facilities are usually voice processing banks that hold innumerable information recordings. These could include weather, traffic information, horoscopes, sports lines, tax advice, etc. Other uses for 700 services are group applications, such as voice or graphics teleconferencing. As the carrier adds more 700 services, the MTS service has greater value to the customer – even though the MTS service itself does not change. These 700 services are available only to presubscribed customers, giving users a non-price reason to select a particular carrier.

WATS

Beyond these basic MTS services lie the broad realm of WATS services. If a user is making more than one hour of long distance calls a month, chances are there are less expensive WATS programs that will allow it to take advantage of enhanced services and volume discounts.

Indeed, as MTS services become less cost effective, WATS becomes the dominant service for the small- to medium-sized user.

New Access Mechanisms

Many people still associate WATS services with dedicated, high-volume long distance traffic, but WATS services can use a range of innovative access methods with a variety of pricing mechanisms. Available WATS services generally use dedicated access lines, WATS access lines, and switched access to get calls from the customer's premises to the carrier point of presence, although many alternative means of access also are available. The switched access WATS services are the most popular access vehicle for the low end of the market, while the DAL products, particularly T-1, are mainly used by the higher volume users.

Traditional Pricing Abandoned

For the most part, traditional WATS banded pricing is all but gone. In 1989, AT&T adopted a mileage-based schedule for the purposes of rating WATS calls, abandoning the banded custom that dominated both WATS and 800 services for more than twenty years. The move is part of a continuing effort by AT&T to restructure all of its long distance services to make them easier to understand and more growth oriented.

Under the old banded services, customers would sign up for banded lines – lines that could complete calls to specified geographical areas known as "bands," which resemble giant concentric circles. band 0 lines typically covered all intrastate calling, band 1 several neighboring states, band 2 a wider area of coverage, and so on. Higher numbered bands could complete calls to all bands below it – a band 3 could complete all band 3, 2, and 1 calls. Band O calls, however, required a separate line for AT&T due to regulatory reasons. (Importantly, this restriction never truly applied to AT&T's competitors.)

From a user's perspective, these banded lines are quite confusing and uneconomical relative to competing, non-banded products. Banded services typically charge the rate applicable to the line, not the call. If a user calls someone in a neighboring state over a band 5 line, the caller would be charged band 5 rates, not the lower band 1 rates which would be applicable if the call were sent over a band 1 line.

This discrepancy was a boon for AT&T's competitors over the years. Optimizing networks with these services required time-consuming engineering analysis and constant maintenance. MCI, Sprint and others took advantage of this and came out with "virtual-banded" services. These formats give users the same banded structure as

Company	Product	Access Method
AT&T	WATS	WATS access line
	Long Distance	Switched lines
	Pro WATS	Switched lines
	Megacom WATS	Dedicated access lines
MCI	Prism III	WATS access line
	MCI Dial 1	Switched lines
	Daytime Savings	Switched lines
	Prism Plus	Switched lines
	Prism I	Dedicated access lines
	Prism II	Dedicated access lines
Sprint	Dial 1	Switched lines
	Sprint Plus	Switched lines
	Dial 1 WATS	Switched lines
	Ultra WATS	Dedicated access lines

Figure 5.2 Available WATS products from the major carriers.

the old AT&T WATS but charge each call according to the mileage, not the line, traversed. This saves customers a substantial amount of money.

As the market has developed, so have the number of competitive options for long distance service. Nearly all of the newer services have opted for the virtual format, but using new rate structures that have largely mimicked MTS rates – mileage bands rate tables with basic per hour or per minute rates. Customers have demanded simple, easy-to-understand straightforward rate structures, and they are getting them.

As an added sign of uniformity across AT&T products, the mileage bands correspond to those we see in the company's other products. This is also generally true for all competitor WATS products. Since WATS mileage bands coincide with most MTS mileage bands, jumping from MTS to WATS is that much easier.

Like AT&T, most of the competitors are abandoning banded WATS services as well. While the major carriers still tariff banded WATS services, they are not actively selling the products, and are working to migrate present users to other services.

Pricing formats do not vary much from carrier to carrier. Over time, competitive pressures have forced the carriers to adopt similar pricing formats for their WATS services – they are largely discount tables based on MTS rates. Indeed, AT&T's PRO WATS program consists of extended discounts off MTS rates in return for a fixed monthly fee. This "WATS" program is nothing more than a billing construct.

Some WATS services are innovative as they stray from the standard band or range definitions. For instance, some of the smaller resellers around the country differentiate themselves by providing a wider calling area than their competitors, particularly in areas where there are strong communities of interest, such as California and Nevada.

US Sprint, which historically has been the most willing of the larger competitors to deviate from AT&T's standard of pricing, created a new band for its WATS products in 1990 – band 1A. This was created for large metropolitan areas where neighboring states and area codes warranted more competitive pricing plans. US Sprint first launched band 1A pricing in the New York metropolitan area. Band 1A rates were applicable for calls to area codes in New York, Connecticut, New Jersey and parts of Pennsylvania. Other logical areas for similar plans include Chicago, Washington, D.C. and San Francisco/Los Angeles, Calif.

Choosing a Long Distance Service

The similarities between MTS, WATS, and, as will be shown in Chapter 8, virtual network services., makes it easier for users to pick a long distance service. The carriers are creating a simple migration path that strings throughout their long distance services. MTS users grow into WATS users, who grow into multi-location WATS users, who eventually need a virtual network service. Pricing formats among the previously distinct services have been adjusted to make migration from one service to the next simple and easy-to understand. WATS pricing for most of the major carriers looks like a stepped discount table off MTS pricing, with greater discounts awarded to the larger volume customers.

Importantly, in designing the services in such a manner, the carriers are reducing complex analysis that has existed in the past. Users no longer have to struggle over which service is best for a particular customer network. MTS, WATS, and even virtual network services have been molded into a consistent growth path with few options within any given market segment. The carriers hope that, with the limited but clear set of options available, a customer, once committed, will feel assured that he or she has selected the appropriate service, and therefore will be less likely to open a network to re-evaluation. This reduces the risk to any one carrier of losing a customer to a competitor.

For the customer, choosing a long distance provider becomes less of a "which service" decision and more of a "which carrier" choice. For instance, any small business (i.e., one with expenditures in the $50-$2000 per month range) typically has the choice of one WATS service with each carrier. In 1989, there would have been three or four WATS services in this same target market from which to choose. AT&T used to have PRO America I, II, and III. Now these have been consolidated into AT&T PRO WATS. Sprint offers; Dial 1 WATS; MCI has Prism Plus. One service per carrier!

Discount Mania

Despite this consolidation of WATS services, market segmentation still is important. Each segment of the market has different cost expectations and the carriers are not oblivious to this. As customers build larger traffic portfolios, they expect larger discounts. In past years, these higher discounts were provided through separate services. Now, they are merely breaks within a discount schedule. Thus, a carrier can create a discount plan with discount tapers of $0-$500, $500-$2000, and $2000 plus, in order to create pricing levels targeted for each of those market segments. Most carriers have designed their products so that they are transparent to the user as he or she moves from one discount taper to the next.

If there is anything easily misunderstood in the WATS market it is supplemental discount plans. In recent years, the carriers have started offering long-term discount plans that reward a customer for signing a contract for long distance service for a 1-5 year period – the longer the term, the larger the discount. Naturally, the carriers have stipulated hefty penalties for customers that cancel their plans early, but there are sufficient escape mechanisms for companies that want to upgrade their networks to bigger and better plans.

Long-term discount plans are growing in scope. Initially, these plans were designed to provide added discounts just to existing WATS rates. Recently, the carriers have moved to include them with other services to make the entire product line more cohesive. For instance, 800 services and WATS services are being combined under virtual network offerings, creating cross-service discounts. As a result, carriers are rewarding users for bringing all of their traffic, not just the WATS or 800 or travel card traffic.

Users should deliberate carefully before they commit to a long-term discount plan. Certainly, there are cost advantages to having greater discounts today; however, the customer is committing its traffic to one carrier and loses the ability to occasionally put a network up for bid – and the potential to further reduce costs. In the end, the decision is a corporate one. Many companies want to concentrate efforts on creating new applications with their long distance service and do not want to be bothered with reconsidering carriers every year. Many of these firms see the relationship with the carrier as a strategic one – a partnership of sorts in which the customer relies heavily on the carrier for long-term network development and applications engineering. Other companies either do not have the intense need for such a close carrier relationship, or have defined their long distance needs in such a generic manner that they can bid their traffic to the lowest provider. Still others fall between both of these – they do not mind committing their traffic to one carrier for a long term, but have no need for a close strategic network relationship.

Multi-Location Plans

Yet another tactic the carriers use to garner all of a customer's long distance traffic are the multi-location calling plans. These provide users with more than one site with a means to aggregate their traffic to gain further discounts. AT&T has its Multi-Location WATS Calling Plan (MLW); MCI has its Corporate Account Service; and Sprint has Network WATS. Each of these operates in fundamentally the same manner – customers are encouraged to add sites onto the calling plan in order to gain higher discounts based on the overall corporate calling levels. Typically, there is a monthly minimum usage level or a monthly fee associated with each program.

These programs serve as a bridge between the WATS and virtual network services. As user grow in network size and usage, they would migrate from a multi-location WATS plan to a virtual network service.

WATS Features

As with other long distance services, pricing differentials in the 30 percent to 50 percent range are hard to find among similar competitor WATS products. With a few notable exceptions, the days of selling solely on low pricing are largely gone. This places more emphasis on criteria other than the service and its pricing, such as billing, network reliability, customer service, etc. Indeed, options such as enhanced billing, alternate routing, fiber-only routing, and added feature capabilities are being offered to boost the attractiveness of WATS services to customers.

Most carriers offer the basic menu of WATS features – validated and non-validated accounting codes, speed dial, travel cards, and magnetic tape billing:

- **accounting codes.** These allow users to allocate costs among different cost centers by punching in two to four digit codes after dialing the telephone number. The monthly bills break out usage according to each code used, facilitating allocation.

- **speed dial.** These allow users to simply dial three to four digit numbers in lieu of full ten to fourteen digit public network numbers. The network translates the speed dial numbers to their public network counterparts and completes the call.

- **travel cards.** These allow access to the long distance carrier when away from the home or office. These cards are discussed in more detail in Chapter 9.

These features are fairly standard at this point. One big differentiating factor for many WATS customers has been in billing. As telecommunications grows as a strategic resource within companies, cost management increases in importance. Carriers often offer customized billing as part of the initial sale pitch. This important topic is covered in Chapter 13.

As with MTS, some of the carriers are looking at offering the "700" services mentioned earlier to increase the customer loyalty to WATS services. Thus, if a customer subscribes to MCI Prism Plus, then his or her telephone lines would have access to all of MCI's value added services.

The Future of MTS and WATS

The market is heading towards more integrated long distance plans in which, once a manager chooses a long distance carrier, the selection of service is nearly automatic. Choices of access will rely more on what enhanced features and further economies the company wishes to accrue.

At some point, the distinctions between virtual network services and WATS services will also blur. In fact, WATS could very well find itself pushed out of the market like MTS as virtual network services are targeted toward lower segments of the marketplace. Right now, long distance carriers are offering customers with usage as low as $3,000 per month virtual network solutions. The carriers are also discussing residential virtual network services. Such residential plans could provide a pricing umbrella for a family's home, grandparents, kids at college – even calling from the road. Throw in a personal 800 number and cellular service and one has a recipe for future personal telephony.

Consider some of the following vertical market oriented products that will address specific niche needs:

1. **Yuppie WATS.** Standard WATS pricing with additional discounts for travel card usage and accounting codes. Professionals in the 21-30 age group are often unmarried and travel frequently for work. While corporations often pick up the tab for calling card traffic, they may not for personal calls. And many young professionals are starting their own businesses and need additional calling card discounts and feature flexibility. Accounting codes are also great for homes where 4-5 professionals share one roof and have problems dividing monthly bills. The carrier could even throw in speed dial features.

2. **Family WATS.** A multi-location WATS package that centralizes billing into the main family account, and covers grandma's house, the kids in college, and any other extended family that the family may be providing for. A personal 800 number could be tied into this account for traveling away from home with applicable cross-service discount schemes. New carrier voice messaging and signaling equipment could be used to notify the household when calling reaches certain thresholds – i.e., when outbound calling tops $50, or 60 minutes.

3. **Commuter package.** WATS service married with cellular service. This would be designed for persons that travel long distances to work every day. Special pricing plans could link cellular usage and long distance usage where applicable, and provide the customer with enhanced services designed for specialized commuter problems. Many cellular providers are already offering such bonus services as "#JAM", which helps users get around traffic jams, and "#LOST", which tells users how to get where they want to go.

4. Point to Point WATS. A WATS service that rewards two sites that subscribe to the same carrier by giving them discounts for calling between the two points. This service is great for long distance relationships, small companies with just two sites, etc.

Some of the carriers are already offering portions of the above services, but complete packages that integrate service, features, and billing are on the way. In the end, pricing will be less of a major factor and the enhancements that come with the product will take on greater proportions in the buying decision.

Chapter 6

800 Services

Who could have predicted that 800 services would grow to be such a fact of life that a company without a toll-free number now is considered to be in the dark ages? Since AT&T first introduced the service in 1967 with some 650 toll-free numbers, toll-free numbers have grown to be an integral part of our economy. Toll-free services add an extra $7 billion to the nation's gross national product each year.

No one has to sell the average company on the benefits of an 800 number. The applications are numerous – order entry, sales, customer service hotlines, etc. Although many companies complained in the past that the service was too expensive, the monthly costs of having an 800 line have dropped to as low as $2.75 per month, lower than many optional local calling features, such as call waiting or call forwarding. In truth, most companies have to look to such items as fraud and abuse of authorization codes, before they can blemish toll-free services. Even then, these problems are so controllable and infrequent that toll-free numbers practically sell themselves.

The public is now well accustomed to 800 numbers. Studies in 1989 showed that more than 50 percent of all consumers used an 800 number at least once in the preceding 12 months. Nearly 20 percent of those surveyed dialed toll-free numbers more than three times in that same period. This trend in 800 calling is only going to continue. Consider this:

Figure 6.1 Toll-free services allowed US residents quick access to information on completing questionaires and obtaining forms for the recent 1990 Census. MCI provided nationwide circuits for calling 1-800-999-1990. Also, there were special numbers for hearing-impaired persons, and for speakers of seven different foreign languages – Spanish, Chinese, Thai, Korean, Cambodian, Vietnamese, and Laotian.

- The market for 800 services is rapidly expanding. While toll-free services used to be just for the nation's largest users, residential consumers are installing 800 numbers in their homes just to cut the costs of kids calling home from college or from the road. Telecom*USA pioneered the concept of residential toll-free services with its launch of a residential 800 service in 1989 with a fixed cost of only $2.75 per month, plus a usage fee of 29 cents per minute.

- Retail market observers predict that the number of retail in-store purchases will continue to drop as shopping at home becomes more popular. Some analysts predict that before the turn of the century almost one third of all purchases will take place away from the store.

- The advent of sophisticated network signaling technology and enhanced network intelligence is turning the "plain vanilla" toll-free services into technology-packed strategic marketing tools. Companies can now terminate 800 calls at different locations based on the calling number, the time of day, or the day of week or year – even how many calls have gone to other corporate locations. Add to that the ability of these services to prompt callers for information or to block certain users from calling are examples of 800 as a powerful corporate tool.

Just how effective can an 800 number be? In 1981, the Quaker Oats Company launched a marketing campaign based around an 800-number for its Cap'n Crunch cereal. More than 24 million callers phoned into its 800-number, stimulating sales of more than 18 million boxes of cereal. Market share during the promotion jumped 33

Company	Product	Type of Access
AT&T	Megacom 800 Service	Dedicated access
	800 Service	WATS line
	Readyline 800 Service	Local line
MCI	MCI 800 Service - Dedicated Termination	Dedicated access
	MCI 800 Service - Switched Termination	WATS line
	Business Line 800	Local line
Sprint	Ultra 800	Dedicated access
	Direct 800	Dedicated access
	FONLine 800	Local line

Figure 6.2 Available 800 products.

percent; even after the promotion was over, market share was still up by more than 15 percent.

How damaging can not having an 800 number be? An AT&T-sponsored survey of telephone buying habits performed a few years ago found that three out of four shoppers (75 percent) are more likely to do business with a firm with an 800 number than an otherwise equivalent company without one. With the proliferation of 800 services since that time, that percentage is likely much higher today.

Current Developments in 800

Toll-free services are evolving rapidly. New features, better capabilities, lower prices, and broader forms of access are just a few of the improvements seen in recent years. The nation's largest long distance carriers are also extending the 800 services overseas – England, France, Japan, and beyond. Specifically, however, there are four main areas of change in the 800 services arena: structure, pricing, coverage, and network intelligence.

Banded Services

One notable happening in toll-free services is the demise of the "banded" structure for 800 services. This goes hand-in-hand with similar changes for traditional banded WATS products, discussed in Chapter 5.

In July, 1989, AT&T adopted a new mileage-based schedule for the purposes of rating 800 Service calls. As with the virtually banded WATS pricing, the revamped 800 services price calls according to the actual distance the call travels – not the line over which the call is received. The changes realigns the AT&T product to meet the new competitive pricing formats of MCI, US Sprint and the other carriers. AT&T created six mileage bands that are now used for all of AT&T's 800 Services. Again, the mileage bands correspond to the same mileage bands we see in AT&T's other products.

Volume Pricing and Long-term Plans

The changes in structure required commensurate changes in pricing. The industry is moving towards a basic generic pricing format for both inbound and outbound voice services – one with mileage-based usage pricing, volume discounts for bulk usage, and long-term plans for added discounts.

In the past, toll-free services had volume discount plans built into the rate tables themselves. These were in the form of tapered rates, i.e. one set of rates for users whose usage averaged between 0 and 25 hours per line, another set for a higher average usage of 25 to 50 hours per line, etc.

Tapered plans can cause confusion on the monthly bill. Call detail can list different call costs for two identical calls placed at different times of the month, depending on the time of the month that the call was made. The calls are priced as they are placed – once a taper point is reached, later calls are billed at the lower taper rates. For the most part, these rate formats have fallen out of style and if they still exist for any particular service it is largely indicative of an outmoded billing system rather than distinctive pricing. For example, US Sprint maintained tapered rates for its Direct 800 and FONLine 800 services because its limited billing system. Not until its Invoice Processing System (IPS) was introduced in 1990 was the carrier was able to adjust the formats.

In place of tapered rating systems, carriers instituted simplified rate plans plus stand-alone volume discount plans. Most have considered or are considering long-term pricing options as well. For discount plans, carriers are opting for the same simple percentage discount tables seen in WATS. These provide different levels of discounts off the total bill, based on that month's usage amount.

Each of the major carriers also has supplemental discount plans for its 800 services. As with the WATS plans, these 800 plans are designed to lock customers into long term agreements in return for greater discounts and may cover more than just 800 services. For instance, AT&T's discount plans started out covering just interstate standard 800 traffic, but grew to include interstate Megacom and 800 Readyline traffic, as well as intrastate traffic on all services. AT&T plans to link 800 services with its virtual network service, Software Defined Network, in a discount plan. As of August, 1990 AT&T's plans for this had been rejected by the FCC. AT&T's next step is still undefined.

Some of the carriers other than AT&T, MCI, and Sprint are more flexible in their pricing packages, as most are not subject to tariffs at the FCC. Multi-year contracts and volume discount plans in these instances are typically negotiated on a client-specific basis. Regional carriers historically have been quite competitive in price, winning business away from the national carriers by appealing to smaller customers who have "standard" 800 service needs.

How An Interstate 800 Call Is Placed

The routing of a toll-free call shows how advanced the public network has become. Toll-free routing requires intelligent network software and databases to make sure the call goes to the right place over the right path. When a caller dials an 800 number, the dialed digits pass over the access line and are received by the serving central office of the user's local telephone company. That office screens the first digits of the call and automatically recognizes it as a toll-free number.

The central office then decides which carrier is assigned to that telephone number by looking up the NXX in its database of carriers. With that information in hand, the local exchange switch forwards the call, with the originating telephone number and other network information, on to the carrier point of presence.

Note that this process varies slightly if the 800 call originates from a customer's dedicated facilities which send calls directly to a carrier POP. In these cases, the carrier performs the screening function and keeps all of its own NXX-defined toll-free calls, and releasing all others into the public network for completion as described above.

Once at the carrier POP, the intelligent network picks up the call and subjects it to a number of tests for routing and special handling. The switch reads the dialed digits and the originating number information and performs a database scan (a "dip" in industry lingo) to read any customer-specific instructions for this call. The most important piece of information is where to send the call. The carrier databases contain translation information regarding the standard public network equivalent of the 800 number. For instance, 1-800-326-6927 might correspond to 1-203-645-0471. This number could change according to customer-specific instructions for different routing by the time of day, the day of the week, originating area codes, and originating states. These special options are described in more detail later in this chapter.

With the handling instructions, the 800 number, now a plain old telephone service (POTS) number, is sent through the network to the specified termination point. This termination point is the customer site and termination options include dedicated access, business lines, WATS access lines, or private bypass facilities.

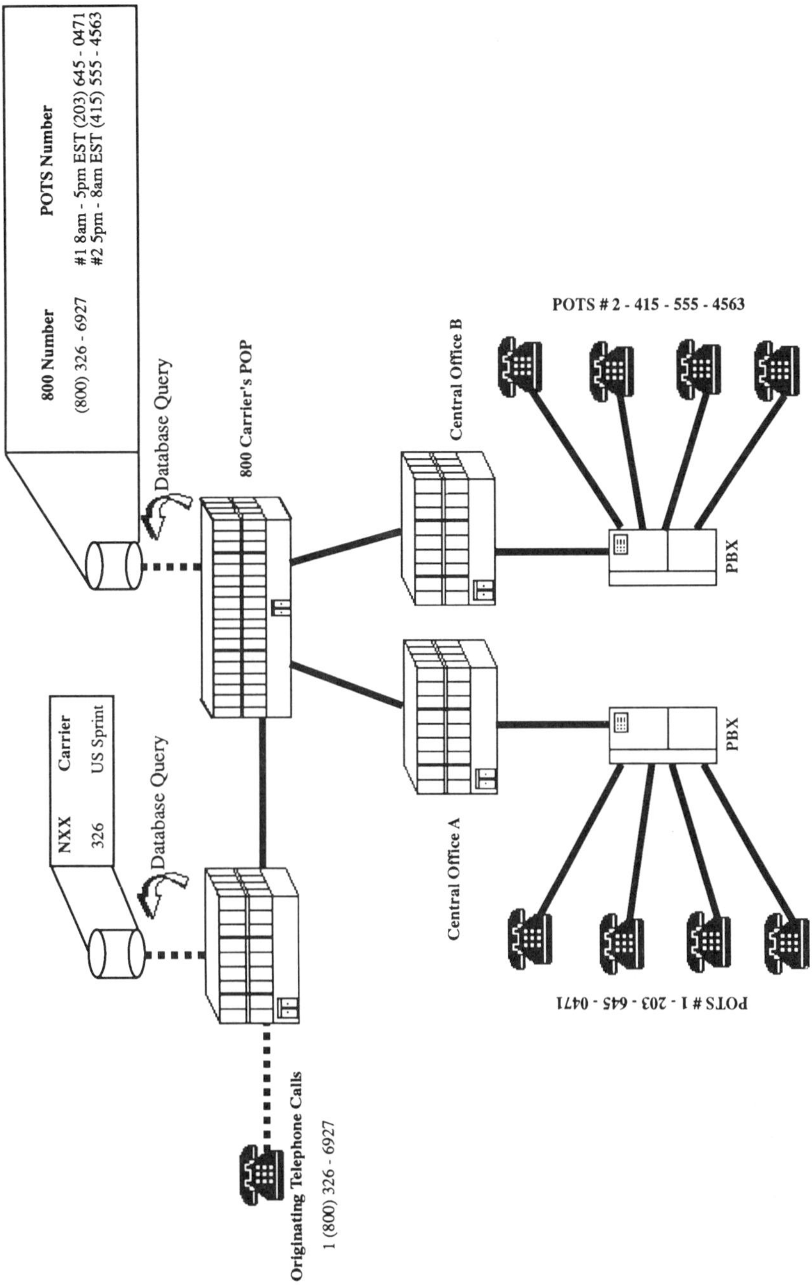

Figure 6.3 The routing of an 800 call requires extensive use of the intelligent network as calls are screened and routed based on information kept in sophisticated databases in local and long distance carrier networks.

Service Band	Day	Evening	Night/ Weekend
1	$9.19	$7.51	$6.26
2	$9.19	$7.91	$6.60
3	$9.95	$8.13	$6.77
4	$10.46	$8.56	$7.13
5	$10.71	$8.77	$7.31
6	$12.25	$10.00	$8.34

Figure 6.4 Sample 800 service pricing

Changing Access Methods

One well publicized change in the 800 industry has been in access methods. The hottest selling 800 services today are the business line-terminated services, such as MCI's Business Line, Sprint's FONLine 800, and AT&T's 800 Readyline.

In the past, toll-free services required dedicated circuits terminating in a PBX. This involved extra ports on the PBX and possibly extra hardware. When the lines could not be used for inbound 800 calls, they typically were not able to be used for anything else, like outbound long distance or local calls.

Times are changing. Access is becoming more economical and more efficient. Business line 800 services use standard local telephone lines to terminate calls so companies don't need extra premise facilities. When there are no inbound 800 calls, the lines can be used for WATS calls, local calls, etc. For the average company, this means reducing the total number of trunks leading to and from a PBX lowering monthly circuit costs.

For many of the same reasons, T-1 is also a hot method of access. Businesses are finding it cheaper to replace analog access facilities with digital T-1 lines for access to WATS, private line, international direct distance dial (IDDD), and other long distance services. Through T-1 access partitioning, toll-free services can piggyback on these circuits as well. This makes more efficient use of a single T-1 circuit, this is more useful to a company's long distance needs. This is especially true with ISDN call-by-call service which can allocate, and reallocate channels depending on the need for them to be used by different applications.

Finally, WATS access line 800 services are being emphasized less by the carriers. In most states, changes in switched access rates (a component of WAL costs) have forced the carriers to emphasize the dedicated access line based and T-1 based services over the WAL based options. However, in select states, such as Texas and California, where WATS access line rates are lower than those for dedicated facilities, carriers push the WAL based options.

800 Coverage Expansion

Toll-free services are expanding in coverage as well. While AT&T, MCI, and Sprint are working to provide worldwide inbound services to the US, small regional domestic carriers are expanding their products to cover larger sections of the country. Carriers such as Litel, Microtel, RCI and others are working to grow outside of their sectional markets. In some cases, the toll-free expansion represents a different focus for the carrier. Microtel, for instance, is looking to launch an in-state Texas 800 service, although all of its other switched services historically have been confined to the Florida and Georgia region. By the end of 1990, there will be a number of new competing national 800 services in place. With more carriers moving into larger markets, the typical end user will have more choice. Especially in large states such as Texas, California, and New York where as many as ten carriers could be providing their own 800 service.

Network Intelligence

Of the changes occurring in the toll-free market, network intelligence has the most sweeping impact on the 800 services. The long distance carrier networks used to be driven largely by hardware – switches, junction boxes, wire, etc. However, as com-

A Look at 800 Features

Toll-free services have been transformed from simple delivery vehicles for calls to full-featured intelligent services with a wide range of applications. Among the more popular features now available from the carriers are:

- **800 Area Code Routing.** Permits the identification of termination points for calls originating from specific geographic areas, such as specific area codes (NPAs).
- **Band Advance.** Permits the rerouting of a call from a group of lines, service group, that has exceeded its capacity to a different 800 service group which encompasses a greater calling area. This typically applies only to the old fashioned 800 services that use traditional banded pricing.
- **800 Call Allocator.** Permits customers to allocate calls between two or more locations by determining the percentage of calls each location should receive.
- **Call Attempt Profile.** A summary report that provides detailed information on attempted calls to an 800 number. Data includes number of calls attempted by a NPA, time of day, and date of call.
- **Call Intercept.** Permits 800 calls to be answered by a recording in the network.
- **800 Call Prompter.** Permits the caller to listen to a recorded message and enter additional digits for routing to the location of their choice.
- **800 Command Routing** (Also called **Alternate Routing** by MCI and **Command Routing** by US Sprint). Permits the customer to access preprogrammed rerouting for 800 calls for use in emergency situations such as downed lines. This alternate routing will be implemented at any time on command from the customer.
- **800 Courtesy Response.** Permits the customer to have calls to an 800 number routed to an intercept announcement on the network.

Continued on next page

A Look at 800 Features (cont'd)

- **800 Day Manager** (Also called **Day of Week Routing** by MCI and US Sprint). Permits calls to the same 800 number to be routed differently depending on the day of the week. This is different from 800 time and day routing in that all calls are rerouted, not just ones from a specific NPA.
- **Dialed Number Identification.** Permits, through the forwarding by the carrier of a 1-7 digit code to the customer facility, the customer to identify the incoming 800 line in a multiple trunk service group.
- **Origination Call Identification.** Permits the customer to receive originating automatic number identification (ANI) data from the network at the time a call is received.
- **800 Referral Service.** Permits the customer to have calls routed to an intercept announcement which refers the caller to an 800 number provided by another carrier when the customer has requested a disconnect with the carrier.
- **Route Advance** (Also called **Direct Termination Overflow** by MCI). Permits the re-routing of a call over business lines or other WATS lines when all dedicated 800 circuits are busy.
- **800 Routing Control Service.** Permits the customer to change the routing plan on a real-time basis from a terminal on their premises.
- **Single Number Service.** Permits the customer to advertise one 800 number for access to multiple interstate and intrastate locations, as well as to multiple groups of lines at the same location.
- **Standard Service Area Selection.** Permits the customer to choose the service area coverage, defined in mileage bands, that he wants. An area 3 coverage would cover service areas 3, 2, 1 and 0.
- **Tailored Call Coverage** (Also **Call Area Code Selection** by US Sprint). Permits the customer to designate regions from

Continued on next page

A Look at 800 Features (cont'd)

which calls may originate. These areas may be as wide as a band, or as small as groups of NPAs.

- **800 Time and Day Routing**. Permits a customer to change the destination of a call originating from one or more NPAs at different times of the day and on different days of the week, according to a specific schedule set up by the customer. This is different from Time and Day manager in that the schedules are defined on an NPA basis.
- **800 Time Manager** (also **Called Time Interval Routing** by MCI and **Time Of Day Routing** by US Sprint.). Permits calls to the same 800 number to be routed differently depending on time of day. This is different from 800 Time and Day Routing in that all calls are rerouted, not just ones from a specific NPA.
- **Uniform Call Distribution**. Permits the user to evenly distribute inbound calls across a service group of 800 lines.
- **800 Validator Service**. Used primarily for credit card validation, this service forgoes the minimum average time requirement on other 800 services.
- **Automatic Number Identification (ANI)**. Permits a customer to receive originating number information from the network at the time the call is placed.

For the end user, these blossoming features can make it hard to compare 800 services, due to the different terms and network technologies among the carriers. AT&T and US Sprint tend to use many of the same terms in defining their advanced 800 service features, but MCI typically calls the functions by another name. For instance, AT&T and US Sprint have Area Code Routing, while MCI has Point-of-Call Routing. All three features provide the user with the means to route calls differently according to the area code from which the toll-free call originates.

puter technology has become more sophisticated, so have the carrier networks. Increasingly, networks are software-controlled, manipulated by programs that control the way they operate and route calls. Network intelligence allows the carriers to identify a call as destined for a particular company, and bound to a particular set of calling restrictions and instructions assigned to the account. Such routing instructions could include time of day routing, numbers to be blocked, and the like. Carriers can add greater functionality by simply installing new computer programs. Network intelligence also drives the carriers' advanced 800 service feature options. AT&T has had more time to develop many of its sophisticated network options. Not surprisingly, the carrier's advanced feature offerings were the most complete in the industry in the first five years after divestiture. However, the competing carriers quickly caught up. MCI, for instance, has been stressing advanced feature and intelligent network development for all of its services as a major part of its overall corporate strategy. Its 800 products now offer a range of features comparable to AT&T's, with the exception of a number of specialized options available from the larger company that target the nation's largest users. MCI is developing a system to allow customer access to its 800 services, similar to AT&T's routing control service, so that users can dynamically control 800 circuits in conjunction with other MCI products. Also, MCI has added voice processing equipment to its network that will enable 800 calls to be intercepted with routing-specific customer-specific announcements.

US Sprint has completed many internal priorities, such as constructing its fiber optic digital network and deploying SS7, and is now concentrating its efforts on adding feature functionality to its 800 services. The carrier has implemented many new 800 service features in 1989 and 1990, aimed at catching up with AT&T and MCI, and has scheduled another series of feature enhancements to its 800 services through 1991 to round out its toll-free product line.

The regional carriers are just starting to enhance their products with extra features. Most of them are concentrating more on simply expanding their market area rather than increasing the functionality of their services. Of the basic features characteristic of the regional services, single number service – being able to have the same number for interstate and intrastate service – is the most common. These regional 800 services will likely appeal to the lower end of the user market where needs for inbound toll-free calling are not feature-intensive.

Database Access

The biggest network intelligence issue revolves around access to number translation databases that allow an 800 number to be translated into a standard 10-digit telephone number for call completion. When AT&T first started providing its 800 services, the carrier developed a proprietary 800 "INWATS" database that stores each 800 number and a corresponding public telephone number to be used to route a call through its long distance network. But AT&T will not let the local telephone companies or the other long distance competitors use that database, leaving the local telephone companies in a predicament. To provide equal toll-free access to AT&T's competitors, 800 numbers by these competitors had to be translated into public 10-digit numbers for completion. The local telephone companies were faced with the problem of figuring out which 800 number should be sent to which carrier for completion. AT&T's competitors developed their own databases for this when calls came directly to them over dedicated facilities, but there was no way for the local telephone companies to do 800 number translations. Eventually the FCC ordered the Bell operating companies to develop their own databases, driven by the SS7 signaling protocol. These new databases are referred to as the Line Information Databases (LIDBs). However, until that system is in place (sometime late in 1991), the BOCs are using a much simpler screening mechanism to route 800 calls.

Under this interim system, each long distance carrier is given a range of NXX's to sell. NXX represents the first three digits of the 7 digit number that is dialed after the 800. For example to call an 800 number a customer dials 1-800-NXX-XXXX, where N is any number between 2 and 9 and X is any number between 0 and 9. When a toll-free call is placed, the BOCs screen the first six digits of any 800 call (i.e., 800-888), look up the carrier for that NXX, and then forward the call to that carrier's network. The long distance carrier performs the 800-number-to-10-digit-number translation and handles the call from there. The problem with this method is that it locks in all NXX's with one or another particular carrier. AT&T customers cannot use MCI without changing their NNX to one on MCI's list and vice versa. If a company wants to change carriers it will have to get an entirely new 800 number.

Number Portability Tied to SS7

The LIDB system does not require Signaling System 7, but does operate more efficiently when it has access to the SS7 protocol. Hopes for an early move to the LIDB system were dampened in April 1989 when the FCC adopted rules tying the transi-

Searching For Vanity Toll-Free Numbers

Many companies use their 800 number to reflect their product or business, such as 1-800-HOLIDAY for Holiday Inn reservations. Vanity numbers may be inconvenient to dial for some users because of the letter-to-number translation which must occur. This is particularly true with some of the newest fancy phones that do not have the letters on the telephone keypad. However, vanity numbers are much easier to remember – especially when on the road and away from a telephone book. Here is a listing of the 800 numbers assigned to the three major carriers – AT&T, MCI, and US Sprint, so readers can see which carrier may have their letter combination. To check number availability, call the carrier.:

AT&T:

221 222 223 225 227 233 237 247 248
255 262 322 323 327 331 332 334 338
342 345 346 356 367 422 423 426 433
443 445 451 458 521 522 523 527 533
537 541 542 543 544 548 553 622 624
634 635 637 638 642 648 654 722 752
782 822 826 832 843 852 872 922 942
962 992

MCI:

234 274 283 284 288 289 333 365 444
456 627 666 678 727 759 777 825 873
876 888 937 950 999

US Sprint

326 347 359 366 669 726 729 733 735
736 767 776 827 829 869 877

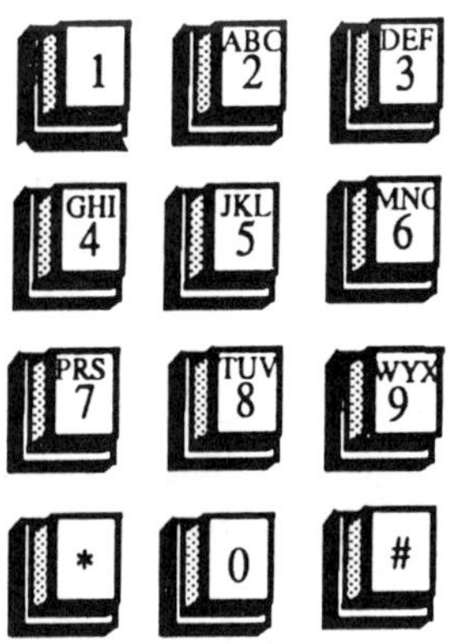

Notes: AT&T: NXXs ending with the digit "2" cannot be used for calls originating from Canada.
MCI: The codes 283 and 937 are reserved for international use, 950 is reserved for the MCI Card.
Other: There have been other allocations of 800 numbers by Bellcore

Here's how some companies are using 800 vanity numbers:

1-800-CLUBINFO	Diners Club
1-800-HILTONS	Hilton Hotel Reservations
1-800-ALLNUTS	Peterson Nut Company

Figure 6.5 800 numbers can help companies sell services before anyone picks up the phone.

tion with deployment of SS7 in the local telephone network. The commission ordered the Bell companies to maintain the 800-NNX access option until they deployed SS7 in all access tandem switches and in end offices serving more than 80 percent of their originating toll-free traffic.

While the BOCs can still implement the new database systems, the FCC's ruling will deter the Bell companies from forcibly migrating long distance carriers from the 800-NXX system. Many will likely still use the existing NXX plan.

The new database system will take longer to process a call. The move to the database without SS7 was squelched because call set-up times would increase from the present two seconds to as much as nine seconds. Implementation under a BOC-installed SS7 system would cut total call set-up time to less than five seconds.

How long will you have to wait? Full SS7 implementation is not expected until 1992.

Selling against AT&T

Until the LIDB database is in place, users are either stuck with their long distance carrier for 800 service or forced to change numbers. However, changing your 800 number may not be as hard as one might think, especially if it is an internal corporate number.

AT&T's competitors have developed programs that would ease the transition from one number to the next. MCI and US Sprint both have 800 number promotional programs that enable a user to efficiently announce the implementation of a new number. Among the items offered by the promotional programs are:

- Rolodex cards that press out of a self-mailing postcard holder that can replace existing business cards or handwritten rolodex cards of your clients.

- Stickers that easily peel off of self-mailing postcards that may be applied to telephones, desk calendars, file folders, etc.

- General purpose stickers—standard or custom sizes—that can be sent in volume to your divisions or outside sales offices.

- Special mailers that consist of specially-designed letters and envelopes—letting people know that important 800 number news is enclosed.

- Multipurpose cards of varying sizes that may be inserted with other mailings to clients, letting them know of your new number.

- A company-specific option that will put your company logos on the cards and stickers.

Prices are nominal. Companies can get 500 rolodex cards, for instance, at a cost of around $25.00 plus postage. Labels can run anywhere from .5 to 3 cents each, depending on quantity ordered and the size of the stickers.

AT&T Call Referral

In wooing customers from AT&T, competitors are aided by an AT&T 800-Service feature, Call Referral. AT&T was ordered by the FCC to provide a recording on its network that states that a number has been changed, and give the new number. This is similar to the recording the local telephone company provides when you move across town.

The competition, however, is somewhat hindered by the fact that AT&T only provides call referral on its MEGACOM 800 and standard 800 products – 800 Readyline customers are left to inform customers themselves if they choose to switch. This favors AT&T in the high-growth business line terminated 800 services niche.

AT&T is very protective of its 800 Readyline client base. The 800 Readyline service has proven to be a phenomenal success for AT&T, much larger than it anticipated in early marketing plans. The changes in switched access charges over the past few years have also made that 800 Readyline more economical than standard 800 in most uses. MCI and Sprint's 800 business line products, like 800 Readyline offer most of the same features as their dedicated 800 products, and those features are growing. MCI and Sprint have steadily implemented new service features over the past year.

Toll-Free and Automatic Number Identification

One controversial issue surrounding the new network intelligence is automatic number identification (ANI) – the ability for the person receiving a call to know from which number the call originates.

In any network, such caller identification information is routinely sent through the network for billing purposes and network management. For subscribers to automatic number identification, this network information is forwarded to the terminating customer location. For commercial users, this is an invaluable tool for capturing critical information about their 800 callers at the time the call. Standard toll-free call detail, provided with bills each month, will give customers the number of the calling party, but this could be a month after the call was completed. AT&T is combating this with on-line call detail. With a modem and a PC, customers can retrieve individual call detail on a real-time basis.

With real-time transmission of this information, a inbound telemarketing center can receive the calling number at the time the call is answered, cross reference this number with a customer database, and call up an information file on the customer just as the attendant is saying hello. For those customers who have called before, new orders may be placed without having to provide the same name, address, etc. information as provided before. For many telemarketers, ANI has cut 45-60 seconds off call times. This translates into clear savings.

ANI has stirred up controversy over the right of privacy callers have to their phone numbers. Is this confidential or public information? Some argue that the phone company has no right to release this information without prior approval; others say that it is a precondition to using the public network. State commissions are wrestling with the issue now and some are considering legislation to bar calling number identification.

The Drawbacks to 800

While 800 has all this and more to offer, there are a number of drawbacks to buying 800 service that few people really consider:

1. **Toll-free lines are the subject of frequent abuse.** Employees tell friends, relatives, even strangers about the new "free" way to reach them at work. While this may certainly cut down on their home telephone bill, the corporate 800 bill will go up. These calls are hard to detect when calls come over standard business lines. However, such abuse may be curbed, through either PBX limitations, selective screening of calls, or use of call information.

2. **Companies are charged for all toll-free calls made, regardless of who is calling.** This includes calls for wrong numbers. One publishing company was wrongly charged for hundreds of dollars of 800 usage when a California bank mistakenly printed the incorrect number for customer service in a

mailing. While the bank reimbursed the publisher for the calls, no compensation was made for personnel time or lost sales due to busy lines. The problem with the newer business line terminated 800 services is that a company never knows when a toll-free call is misdialed – the caller usually just hangs up.

3. It is hard to tell when an 800 number is out of service. Large companies usually find out from the drastic decrease in calls, but smaller companies have a more difficult time. It is especially hard when calls terminate over standard business lines. Until someone complains, many companies are unaware of outages.

These are the major problems with 800 services. However, with some supervision, most of these may be minimized.

The Future of 800

Toll-free services will continue to grow in the future. The explosion of advanced feature development and applications will continue, if not accelerate. Look also for these developments:

- Integration of multiple 800 numbers serving multiple locations into a single number controlled by advanced prompting messages and routing features. Remote management and call tracking will enable companies to use the "combined" network to boost capacity to handle peak demand situations, such as television or radio promotions.

- High growth in the "900" services. Firms will look to a combination of 800 and 900 services to fulfill corporate communications needs. The software company of tomorrow, for instance, could offer its patrons free software support for the first 30 days via an 800 number, and charged support thereafter via a 900 number. The burden will be placed on the carriers to provide better solutions for peak demand conditions on the network than are presently available.

- Increased consumption of 800 numbers by lawyers, brokers, accountants, voice mail firms, and paging companies as they seek to isolate and billback customers for use of 800 numbers. By providing each client with their own 800 number, companies can track that client's usage more precisely than with call distributors. Voice mail and paging companies could take advantage of improvements in technology – a caller could dial an 800 number into a voice mail firm, connect to a voice mailbox, have the person called paged, and then be cross-connected with that person for live two-way communication. Demand for 800 numbers among these groups is likely to rapidly cut into the available supply of 800 numbers.

- The maturity of the network-premise interface, providing an intelligent linkage between the carrier networks and customer premises equipment. In the near future, customers will be able to dynamically assign bandwidth to different services through customer premise based terminals. Automatic network routing plans could be implemented automatically, based on network conditions such as busy circuits, closed offices, and down lines. Toll-free networks will be automatically driven by software, giving users greater control over where calls are going to and when they are to be delivered to particular sites. To some extent, a few of the carriers provide these services today.

While 800 numbers were a novelty item for a company a decade ago, they will be a fact of life for the company of the 1990s. For the vast majority of users, the question is not whether or not to get an 800 number, it is when.

Chapter 7

Private Line Services

As with long distance services, private line services do not look the same anymore. The boundaries delineating the different forms of private line services have become increasingly blurred recently, making it hard to decide which service to buy. The implementation of digital transmission systems and enhanced routing technology in carrier networks has changed the underlying technical and cost assumptions for the carriers – prompting them to de-emphasize or abolish certain private line services while creating others.

Many of the changes can be attributed to the massive fiber construction programs started in the mid-1980's and still going on today. The resulting amount of bandwidth capacity that was dumped on the market prompted fire sale prices to win market share and fill the network trunks. The effects of these capitalization programs are still rippling through the market, and T-1 circuit prices of 40 percent to 60 percent below AT&T's rates are not altogether uncommon. The glut of capacity has pushed T-1 prices down so far that the lower capacity, lower quality voice grade services may be squeezed out in the near future.

For the private line manager with applications to support, this news appears to be good – more services, faster speeds, and lower prices. The question is, where will it go from here?

The Basics of T-1

The term T-1 belongs to a group of transmission systems referred to as "T-carrier" systems. A T-carrier is a digital system standardized by AT&T for short- to long-distance transmissions. T-carriers generally operate at an aggregate data rate of 1.544 Mbps or greater.

The terms "T-1" and "DS-1" are often the source of much confusion. Technically, T-1 is a type of transmission system and DS-1 is a type of signal. Thus, a DS-1 signal would travel over a T-1 link. The DS-1 rate, according to the North American standard, is 1.544 Mbps. However, foreign carriers have standardized on 2.048 Mbps, generally referred to as "E-1."

A DS-1 signal consists of 24 channels at 64 kbps each. This totals 1.536 Mbps. The remaining 8 kbps is a control channel that keep the multiplexers at each end of the T-1 link synchronized.

The DS-1 signal is part of a complete digital system (DS) hierarchy that has digital signal level zero (DS-0), a single 64 kbps digital channel, as its foundation. By multiplexing 24 DS-0 channels together, one achieves a DS-1 signal. Four DS-1 circuits may be multiplexed to form a DS-2 circuit, or 6.312 Mbps. Seven DS-2s or 28 DS-1s may be multiplexed to create a DS-3 signal, or 44.6 Mbps. At present, DS-3 is the highest level signal generally available to end users in the public network, although higher levels are available for carrier inter-office transmissions.

T-1 circuits have actually been around for quite awhile. They were first used in the public network for interoffice channels in 1962. They provided voice and data communications over high-capacity routes, useful in areas where transmission pipe capacity was limited.

Laboratory studies concluded that a single two-pair wire could replace 24 interoffice trunks when those 24 voice signals were digitized and multiplexed into a 1.544 Mbps signal. This meant a twelve-fold reduction in circuit space. The 1.544 Mbps signal itself was selected because it represented the highest speed of signal that could reliably travel between manholes in metropolitan areas. Manholes have long served as the primary access points to equipment used in telephone transmission, including cabling, repeaters, couplers, junction boxes, etc. The T-1 specifications were written to match the spacing between manholes – approximately every 6,000 feet.

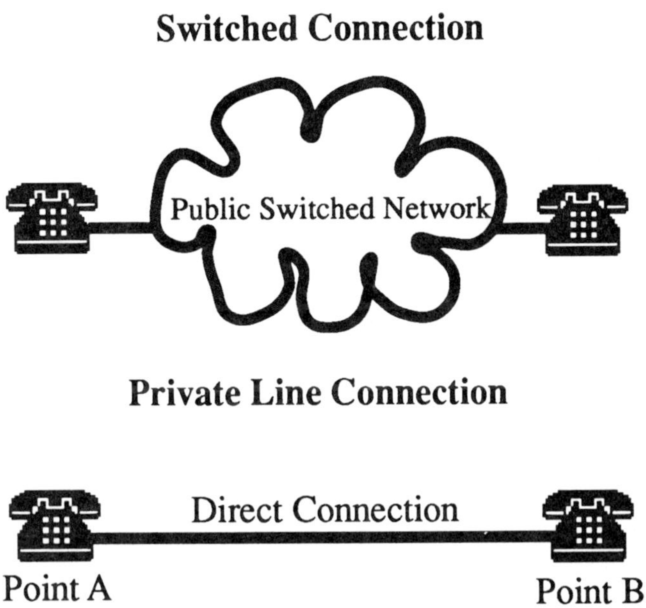

Figure 7.1 A private line connects two sites with a direct channel.

Defining Private Lines

Private lines are nothing more than point to point telephone lines that are directly connected to customer terminal equipment at both ends of the circuit. There are three components to the circuit – the front-end access, the long haul interoffice trunk, and the rear-end egress. The first and last portions are typically provided by the local exchange company, although many private line customers use alternative local access providers or bypass facilities as well. The interoffice channel is provided by the long distance carrier when it is an interexchange private line; intraexchange facilities are provided by the local exchange carrier.

Private line services are no longer limited to the telex and voice grade circuits that dominated the early years of long distance. Now, companies can purchase a full range of both analog and digital products from a wide variety of full service and reseller companies.

Among the services now available are:

- **Voice grade services.** The standard analog private line circuit that has served as the foundation of private line networks for years. The carriers are replacing circuits of this type with digital T-1, fractional T-1, or DDS circuits.

- **Digital data services (DDS).** Digital services typically offering transmission speeds of 2.4, 4.8, 9.6, and 56 kbps. As old equipment is retired, the 2.4 and 4.8 kbps speeds will disappear; most of the emphasis today is on 9.6 and 56 kbps transmission, although the latest switched data services, discussed in Chapter 8, are challenging these too. Lately, clear channel digital 64 kbps digital data services have been introduced.

- **Fractional T-1.** The newest services offering portions of a 1.544 Mbps T-1 circuit for a "fraction" of the cost. The services are targeted for customers who need more digital bandwidth than provided by DDS or switched 56 kbps services but do not want (or cannot afford) a full T-1 circuit.

- **T-1.** Almost the de facto backbone circuit of the network today, T-1 circuits offer 1.544 Mbps of digital transmission capacity. Costs for T-1 circuits have been plummeting in recent years, bringing on price wars reminiscent of those between competitor's WATS services in the early years of divestiture. Significantly, the services are priced low enough to tempt DDS and switched 56 users to convert networks to dedicated T-1-based private line networks.

- **Fractional T-3.** Conceptually the same as fractional T-1 circuits, these higher capacity circuits offering portions of a 44.76 Mbps T-3 circuit for a portion of the cost of a full T-3. These services are not available yet today, but will be soon – a natural market evolution. The services are targeted for customers who need more digital bandwidth than provided through T-1 services but do not want or cannot afford a full T-1 circuit.

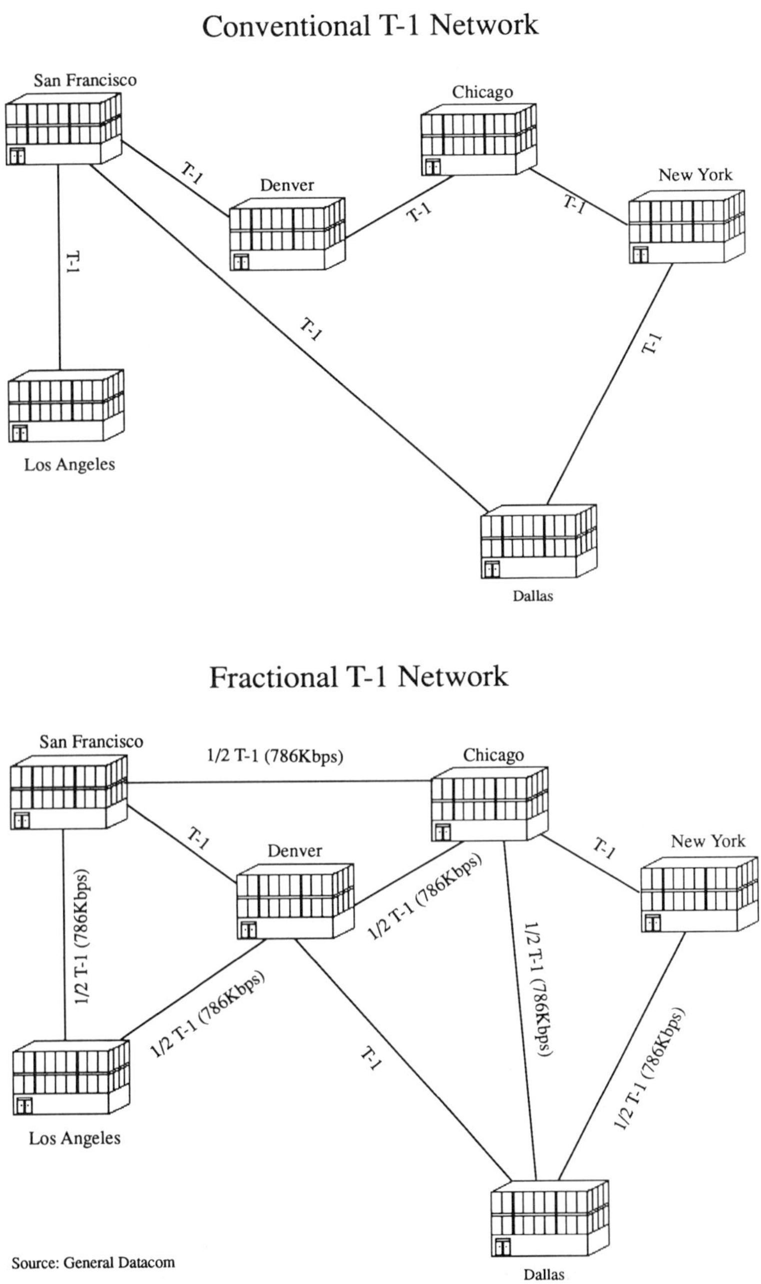

Source: General Datacom

Figure 7.2 Fractional T-1 services, as opposed to conventional T-1 service, allow companies to alternately route private lines to avoid total system failure in the event of an emergency.

- **T-3.** Currently used only by the nation's highest volume customers, T-3 represents 44.76 Mbps of transmission capacity, or 672 multiplexed 64 kbps digital voice and data circuits.

The different carrier networks each have varying levels of availability for the above services. Most of the carriers provide at least one offering in each service category; however, these services may not be available network-wide. For instance, a user cannot get T-1 service from every AT&T central office because the carrier maintains a separate sub-network for its Accunet T1.5 service. Also, some carriers may not be able to provide analog service because of how advanced their networks have become. US Sprint, for instance, has an all digital network. It offers analog private line service, but it is not truly analog. The carrier converts the inbound analog signal to digital for long haul transmission, and reconverts to analog at the far end. Shortly, analog services will become obsolete as users replace analog equipment with more reliable digital facilities. Pricing is also causing dramatic change. As with WATS services, long distance carriers are creating services between services, providing a more discrete migration path for private line services.

Logically, a user would upgrade from voice grade circuits and DDS channels, to fractional T-1 services, to full T-1 links and eventually onto T-3 trunks. However, this migration path is not always so clear cut. In some cases, the pricing of the various services prompts users to skip a level to higher alternatives.

For example, look at T-1 services. Theoretically, 24 voice grade circuits should warrant a cross-over to a high-capacity T-1 circuit that offers a multiplicity of side benefits, such as digital transmission, sub-rate transmission, etc. But in this competitive market, as low as four voice grade circuits can justify installing a T-1 circuit, simply on the basis of monthly circuit charges. With cutovers this low, voice grade circuits may simply cease to exist and be replaced with T-1 offerings. If a company cannot afford a full T-1, then they would likely opt for a fractional T-1.

Fractional T-1

However, low T-1 prices could endanger fractional T-1 services too. Fractional T-1 has been a highly-publicized private line product in the past few years. Although, there have been innumerable articles discussing the economic and network benefits to be derived from the product, it is not clear yet whether the fractional T-1 services are all they are cracked up to be. While conceptually a great idea, the services may have been obsolete almost before they were launched. The continual decrease in the pricing of T-1 services may prompt customers to simply skip to full T-1 products,

Company	Product	Product Type
AT&T	Accunet T45 Service	T-3
	Accunet T1.5 Service	T-1
	Accunet Spectrum of Digital Services (ASDS)	Fractional T-1
	Dataphone Digital Service	Digital data service
	Voice-Grade Private Line	Voice-grade private line
MCI	TDS 45	T-3
	TDS 1.5	T-1
	Digital Private Line Service	Fractional T-1
	Digital Data Service	Digital data service
	Voice-Grade Private Line	Voice-grade private line
Sprint	Clearline 45	T-3
	Clearline 1.5	T-1
	Clearline Fractional T-1	Fractional T-1
	Clearline Digital Data Service	Digital data service
	Clearline Voiceband	Voice-grade private line

Figure 7.3 Available private line offerings.

particularly given the lack of complementary fractional T-1 access on the part of the local exchange carriers. Local access must be through full T-1 circuits. While the long distance carriers rush to market their fractional T-1 services, local exchange carriers are wondering if they want to develop the services at all. For the end user, this limits economic feasibility of installing fractional T-1 links in a network.

IXC Fractional T-1 Products

The fractional T-1 services debuted a few years ago, with Cable & Wireless being among the first to offer the service commercially. AT&T followed in June, 1989 with its Accunet Spectrum of Digital Services (ASDS), and MCI and US Sprint followed in fall, 1989 with similar products.

The largest application for the service likely is for large users who wish to provide more disaster recovery and alternate routing in their networks. By splitting a T-1 into two one-half T-1 circuits, the user can achieve greater security in case of an IXC line failure. Smaller users will also be attracted to the service as fractional T-1 allows them to gain economic efficiencies of bulk transport. The lower costs of the IXC long-haul circuit, relative to existing DDS or voice grade circuits, can offset the higher costs of the full T-1 access. With a high-capacity line to the carrier point of presence, the user then can easily "grow" the fractional T-1 circuit until a full T-1 line is attained. For most small businesses, this is enough growth capacity for the near to long term.

The RBOCs and Fractional T-1

To date, only one RBOC, NYNEX, has launched a fractional T-1 service compatible to the IXC offerings. Of the remaining RBOCs, only one other, US West, has said that it is actively pursuing a fractional T-1 service. Other RBOCs are questioning whether the fractional T-1 services are necessary in today's marketplace. In their minds, full T-1 access is already priced low enough for users that want to interface to IXC fractional T-1 services. With the other private line services available on the local level – DDS, analog voice and data, public switched data, as well as T-1 – many managers in the local exchange carriers believe that even the IXC fractional T-1 services may be unnecessary.

Moreover, economically, the T-1 access may make very little difference in the overall cost of using fractional T-1 service. According to the Aries Group/MPSG in Rockville, MD, under New England Telephone's January 1, 1990 tariffs, a user in downtown Boston would be charged $501.75 for a 12-channel, two-mile fractional

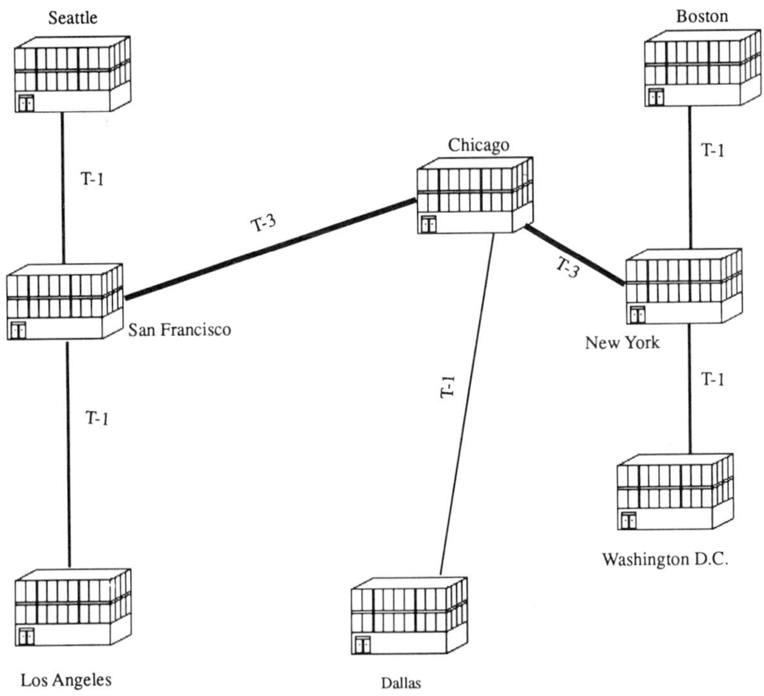

Source: TeleChoice, Inc.

Figure 7.4 T-3 backbone networking allows companies to gain cost economies over the "super highway" routes of the network.

T-1 access circuit; a comparable full 24-channel T-1 link would cost $594.43. To gain twelve more channels, a user would merely have to pay just over $90 – quite a deal. Cost differences only come into play with longer distance access circuits. A twelve-circuit, thirteen-mile fractional T-1 circuit would cost $712 while a T-1 link would be $943.68 – a much greater difference.

Clearly, the local exchange carriers and the IXCs have different motivations for offering fractional T-1 services. The IXCs can gain more revenue by splitting up a full T-1 into smaller increments, because each piece can be sold for more than the commensurate fraction of the full T-1 price. The carrier also will not have a problem filling the capacity on the rest of the circuit because it has hundreds of other fractional customers – not to mention switched traffic – that could fill the gap.

For a local exchange carrier, the operating cost of the line between the customer–premise and the central office is the same, whether it's a full T-1 or a fractional T-1 link. The local exchange carrier would lose revenues if it replaces a full T-1 circuit with a partial circuit. Generally, there is little chance for the local carrier to sell the remaining unused portion of the fractional local access link to other users. The only portion of the circuit in which the local exchange carrier can cut costs is in the shared transport from the customer's serving central office to the carrier point of presence. Combined with the upward migration trend of users to full T-1 circuits, the fractional T-1 market segment may not be large enough for the local exchange carriers to target.

With T-1 prices still dropping, and the lack of interest among many of the local exchange carriers, fractional T-1 may have a short history. With higher bandwidth applications coming down the road, many users may just accept the added expense and buy the full T-1 circuit for the flexibility of the extra capacity.

T-3

The success of T-1 in the marketplace provides a glimpse of what's to come for T-3. As fractional T-1 and DDS users migrate up to T-1, so will T-1 users grow into T-3. The still plummeting costs of this digital capacity is luring even the most conservative companies into exploring high-bandwidth applications such as imaging and video.

The cutover point between T-1 and T-3 is rather low. For 300-mile circuits, for instance, it is economical to move to T-3 if a company is using 10 or more T-1 circuits, according to AT&T's January 1, 1990 pricing.

The largest private line carriers – AT&T, MCI, Sprint, Williams Telecommunications Group, and others – are rapidly preparing for this onslaught of demand. Changes to the network to accommodate higher volume and faster speeds of private line traffic include more sophisticated network management access and greater quality of network transmission. Corporate users are putting a lot of pressure on the industry to get prepared for the upcoming demand, saying, "We don't want it now, but we will, so get ready." T-1 multiplexer vendors are finding that customers are writing T-3 specifications into T-1 multiplexer bids. The desire for advanced applications is driving many of these T-3 mux designs. Most present T-3 multiplexers, called M13s, can handle only point to point transmission. They cannot separate individual T-1s or any portion of a T-1 without first breaking the T-3 into the 28 T-1s that comprise the circuit. This is particularly unsuitable for customers who wish to use switching on extensive private line networks that consist of everything from fractional T-1s up to T-3s. However, the next generation of T-3 products is emerging, which are much more sophisticated and will allow a whole range of options not supported by the present T-3 multiplexers.

Also of significance is the incorporation of the SONET (Synchronous Optical Network) standard in these new muxes. SONET is an international standard for carrying T-3s over fiber that defines DS-3 speed as 51.84 Mbps. It provides for the synchronization of DS-1s within a DS-3 – which is not supported by the current M13 multiplexers. This synchronization allows users to add or drop individual T-1s from a T-3 link. SONET is the international version of the SYNTRAN (Synchronous Transmission) standard that redefined DS-3 as 44.736 Mbps. SYNTRAN is only used in the U.S. and is not widely supported by hardware vendors or carriers.

For managers, these new standards will add flexibility in switching DS-3 digital pipelines. SYNTRAN or SONET based multiplexers will be able to support present DDS or high-end WATS and 800 services that presently use the M13 interface. All that is needed is the appropriate interface card.

These technological changes will allow new services, fractional T-3, for instance, for T-1 users who wish to upgrade to T-3 but cannot yet afford the full T-3 line. With more T-3 lines being installed, the likelihood of such a fractional bridge between the T-1 and T-3 services is quite great, particularly given the next generation of intelligent multiplexers and digital cross-connects which are supposed to support these services.

Switched Private Line Services

Many of the same network improvements that allow carriers to launch enhanced 800 and virtual network features are enabling them to offer switched private line services. These are switched versions of the traditional point to point dedicated services. Historically, private line network managers refused to consider anything but a point to point dedicated facility for data transmission. The managers insisted on the high degree of security and performance resulting from having the “closed” environment of a private line. However, ubiquitous digital connectivity and higher network performance standards have raised the overall performance of the public network, enabling carriers to send data calls over these previously voice only networks. For now, these services act as backup facilities in case the dedicated private lines fail. However, more and more carriers are pushing these products as primary facilities.

What Is A DACS?

DACS is a frequently used term in private line applications, particularly in disaster recovery. It stands for digital access and cross-connect system. Basically, DACS is a switch that is capable of monitoring and routing T-1 channels on a DS-0 basis.

The major feature of these systems is that they are software-controllable. Circuits may be rerouted instantaneously – no longer is a technician required to visit the switch and manaully change the private line wiring to accomplish a change.

DACS provides users with a number of network benefits. First, it allows users to centralize and concentrate private line traffic. For instance, a user might have multiple DS-0 channels coming from several different offices within one LATA and combine these into one T-1 channel to maximize the long-haul transmission economies. Before, separate DS-0 interoffice channels or expensive customer premises equipment might have been required.

Another major application for DACS is customer-controlled reconfiguration, or CCR. The underlying concept of CCR is to give the customer the capability to provision its own T-1 circuits. Before, the carriers were doing this for the customer. Users can enter alternate routing tables into the software that can be implemented whenever there is a disaster in one portion of the customer network. If a circuit fails between a customer site in New York and the headquarters in Chicago, then the DACS could be programmed to bypass those circuits to the stand-by data center in Minneapolis – all with the push of a few buttons.

DACS has revolutionized the way users look at private lines. The flexibility given to both carriers and users cannot be understated. Indeed, DACS now serves as an important tool for achieving bandwidth on demand from the long distance carriers.

These "virtual" private line services are just starting to become available, as part of or as an accompaniment to the virtual network services at AT&T, MCI, and US Sprint. The first services launched were switched 56 kbps data transmission services. Other private line-related services released under the virtual network service um-

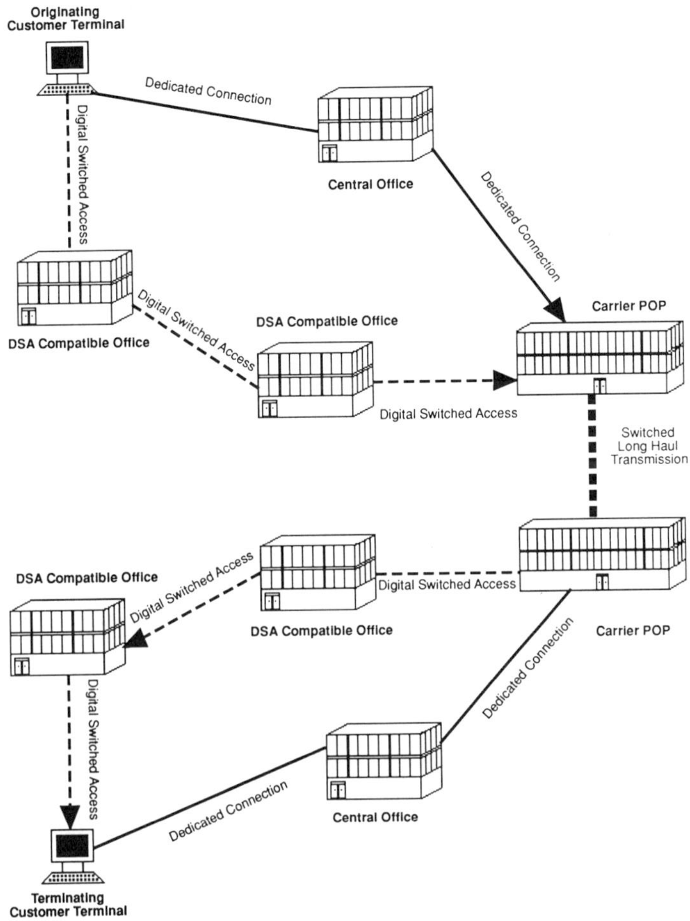

Figure 7.5 Switched reception versus dedicated transmission routes.

brella are largely voice-oriented. One is virtual foreign exchange (FX), which provides the same LATA-wide FX service that is available with dedicated circuits – except the long haul portion is switched. Another service is virtual ringdown circuits, which automatically dial a number when the phone is taken off the hook.

All of these are precursors of an age when dedicated facilities will be unnecessary because network management will isolate and define the parameters of a switched transmission – just as with present dedicated facilities. For users, this will mean greater flexibility and ultimately lower pricing. Users should not be surprised to see switched T-1 service by the first part of 1991 and switched T-3 soon thereafter.

Private Line Enhancements

Private line customer thirst for more enhanced network management information and control is spawning a new era of customer controlled access into carrier databases and support facilities – providing more "hands on" manipulation and testing of private network facilities. The benefits are quicker diagnostic information, faster circuit rerouting, and more control over the entire private network.

Furthermore, soon even more enhanced versions will emerge that cover "virtual" switched services, allowing for dial-up "bandwidth on demand" and total flexibility in managing multiple services over the same network.

Carriers Fight a Mature Market

The natural long-term trend of any market is towards generic products. Right now, many private line services are perilously close to such homogeneity. Private lines have long been seen as components of an overall private network – components that can be easily supplied by a number of vendors. As a result, the nation's long distance providers have been constantly seeking ways to prevent the private line market from being sold on price alone.

Voice grade private lines are a classic example. Analog voice grade circuits won early distinction as a means to cut bulk point to point long distance costs. As greater applications were found for the product, extra features and options were added to give the product a broader level of use. In recent years, however, all of the carrier voice grade offerings have largely gravitated towards one another, making voice grade private lines a mostly generic product. The products have changed little in several years. As applications demanded more stringent parameters and options, other products – digital ones – were created by the carriers. Further, with T-1 prices coming down, voice grade circuits have moved past the level of maturity into obsolescence, as they are replaced by fractional T-1.

The same is true of the digital data services. Most carriers are abandoning them in favor of fractional T-1 and T-1 offerings. AT&T, for instance, is migrating its users to its fractional offering, Accunet Spectrum of Digital Services, as fast as possible. Pricing is becoming prohibitive to stay with DDS.

This leaves the fast growth, profitable digital products – fractional T-1, T-1 and T-3. Competitive pricing is ferocious, and carriers are seeking both to keep their products as "un-generic" as possible to minimize further price wars and to breed customer loyalty.

Customizing Private Lines

As a result, the carriers have started launching several private line related feature and option programs, enhancements that make the private line attractive on the basis of something more than just capacity. Some of these are network-related, such as fiber-only routing and automatic alternate routing. Others are product-specific, such as private line reconfiguration and digital access cross-connect service (DACS).

Most of the private line enhancements that a carrier releases to its customer base are merely extensions of in-house capabilities used to manage and maintain the carrier's own network. Only recently have customer interfaces been developed to allow the end user to participate in these management services.

Private line management offerings are primarily the result of customer-requested applications. AT&T predictably has the most advanced levels of private line management, since its network and private line facilities have been around longer. Some of the newer entrants to "enhanced private line offerings" – notably MCI and Williams – have either launched or are finalizing extensive packages to compete in configuration and maintenance management. In addition, some of the smaller carriers, such as NorLight and ATC/Microtel, are also offering customized access on an individual case basis. US Sprint is a relative newcomer to the private line market, having introduced its first product in 1989, but is busy developing its complementary private line management options.

These product enhancements will segment the marketplace for private lines. In the end, the user will be left with two choices:

1. A manager can choose to keep private lines a generic product and provide his or her own necessary multiplexing, management, and testing facilities. This will increase the user's buying power as private lines can be compared and purchased as simple "capacity pipes."

2. The manager can view a private line service as an enhanced service, recognizing that shopping for enhanced options might limit the number of available providers to few or even a single source. This can increase expenditures but provide greater flexibility.

Network Management Offerings

Private line reconfiguration is one area where carriers are attempting to add extra value to their private line services. Applications include:

- **Network Optimization.** Users can optimize networks with reconfiguration features through either on-demand or scheduled changes to a network's design.

- **Disaster Recovery.** When a point of presence or local central office fails, reconfiguration features can provide rerouting, typically in 5-30 minutes.

- **Alternate Routing.** Preprogrammed alternate paths may be set up for implementation on demand – whether an emergency or standard network need.

Other network management functions will grow from these initial offerings, depending on customer demand.

AT&T's Network Products

AT&T has two main service functions aimed at users of its Accunet T1.5 private line product. The first is Customer Controlled Reconfiguration (CCR). AT&T's CCR service functions allow a customer to electronically reconfigure DS-0 channels within AT&T's Accunet T1.5 circuits via a premises-based terminal. This gives the customer access to AT&T's DACS.

The CCR architecture is rather simple. The customer connects with AT&T's Special Service Management System (SMSS) in Freehold, NJ, either through a private line or dial facility. If a customer dials in, AT&T's computer will dial back the customer device after the security code is entered. This provides an added level of security as the computer is pre-programmed to dial back only specific numbers. This avoids the potential of computer hackers rerouting a user's network. Network-wide DS-0 cross connection information is stored in the SMSS database. Each DS-0 is listed as a "path" through the AT&T network. The customer can store multiple network paths together as a "map" of a desired network configuration. These maps may be invoked on demand or on a scheduled basis. Support is limited to the DS-1 level. Further, CCR is reserved only for private line services – no connections to switched services via M24 multiplexing or direct DS-1 interface is available.

A second network tool for T1.5 users is the Bandwidth Management Service (BMS). BMS provides AT&T customers with rapid reconfiguration and restoration capabilities. It is a management tool that automatically reconfigures a customer's network in less than one minute. It also allows customers to test their networks and monitor network performance. BMS is a feature of AT&T's Accunet T1.5. It uses AT&T multiplexer technology in the serving office to configure and cross-connect DS-0 circuits within and between Accunet T1.5 circuits. Among the many specific capabilities of BMS are:

- Automatic alternate routing of customer cross-connected circuits between nodes.
- Reconfiguration of customer internodal networks.
- Rapid configuration and reconfiguration of Accunet T1.5 networks.
- Reporting of performance monitoring, traffic measurement, alarm information, customer-initiated testing and on-demand selective trouble reporting.

BMS customer's private line networks are routed through a series of points of presences that are controlled by devices called BMS node controllers. These node controllers communicate with every other node controller in the customer's network through in-band data links, maintaining a constant picture of the customer network.

The BMS customer interfaces with the node controllers through a dedicated 9.6 kbps DDS connection between the customer premises and AT&T's 3B2/400 BMS system controller, in Freehold, NJ. That system controller is then linked with the BMS node controllers through network administration ports (NAPs).

The BMS configuration functions can be on a scheduled, unscheduled, or automatic basis. In automatic reconfiguration, the System Controller reacts to facility or equipment failure and, based on a predetermined set of instructions, reconfigures the network around that failure. In scheduled reconfiguration, the customer instructs the System Controller to reconfigure the network according to a preplanned schedule. Customers often use this for time-of-day and day-of-week changes. In unscheduled reconfiguration, the customer uses its remote terminal to instruct the system controller to reconfigure the network on an ad hoc basis.

In 1990, AT&T upgraded BMS by adding its "BMS-Extended," or BMS-E. This provides coordinated end to end control of customer-provided equipment with

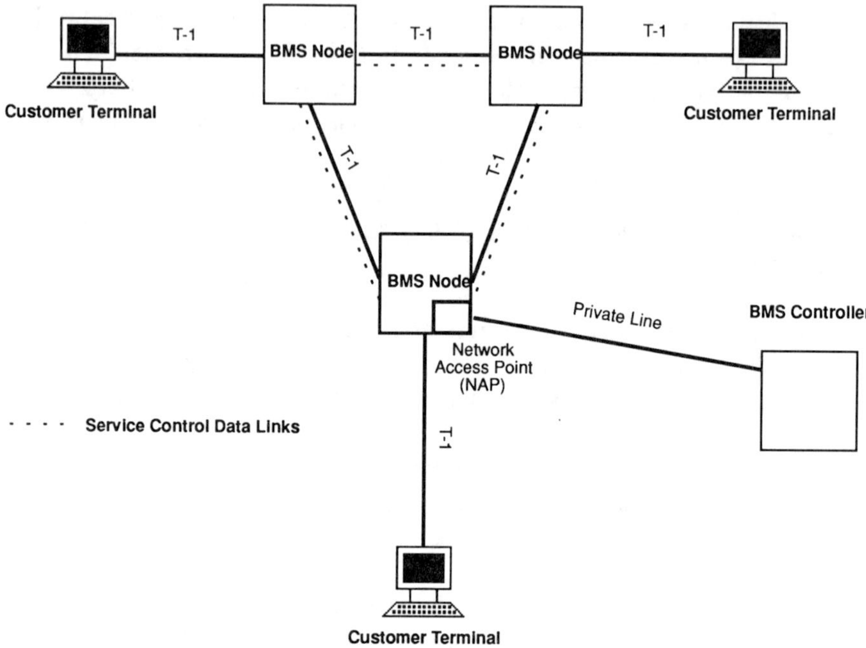

Figure 7.6 Bandwidth managment system (BMS).

AT&T node controllers, including the local access lines. BMS-E uses the supervisory data link (SDL) Protocol that allows control of a network from the customer's premises via SDL-compatible equipment, such as AT&T's Acculink network manager.

Pricing for BMS and BMS-E is considered extremely expensive in the industry. There are nonrecurring and recurring charges for the system controller, node controllers at each AT&T serving office on the customer's network, and a third element, the BMS port. One BMS port is needed for each 1.544 Mbps channel connected to a node controller. To cut the impact on the telecom manager's budget, AT&T offers term and discount plans that include BMS, as well as other office functions such as route diversity and subrate data multiplexing. Under the newest plans, these optional features are discounted at the same rate as the associated intercity channels. This can mean up to 56 percent discounts.

The major differences between BMS and CCR are ones of speed and depth. CCR provides reconfiguration in an average of five minutes for networks, but can take longer. CCR users generally do not require the automatic restoration and network management capabilities of BMS. CCR is targeted for customers with mainly voice grade circuits.

BMS is attractive to Accunet T1.5 customers with large and medium-sized networks who require advanced management features for critical data applications that are integrated with voice grade and video circuits.

MCI's INMS

MCI's private line management options tie in closely with the carrier's overall network management package, marketed as Integrated Network Management Services (INMS). MCI is molding its various individual systems into a integrated management approach for customers with hybrid networks.

MCI's INMS consists of six application groups for network management, one of which is configuration management. MCI has two primary products that comprise configuration management. One is Configuration Manager (previously called Customer Information Manager System or CIMS), which manages virtual network configurations. The other is Digital Reconfiguration Service (DRS), which controls configurations for private lines. The two services allow customers to rearrange their services from premises-based terminals.

DRS represents MCI's first venture into digital configuration management. Its terminal is connected to a host system, which controls MCI's cross-connect switches. Customer changes are implemented in an average of five minutes.

The first phase of DRS, available now, is called "fixed network reconfiguration." This is being used for many of the same applications as AT&T's CCR service:

- **Drop and Insert.** This function is available among the more than 200 cities served by TDS-1.5. It allows a manager to have intermediary sites add data to the T-1 or receive data from it on a backbone network. Drop and insert is accomplished at the DS-0 or full T-1 level on private leased backbone circuits.

- **Load Balancing.** Customers are able to separate circuits for specific applications. DRS allows the customers to allocate application-specific bandwidth among limited resources.

- **Smart Multiplexer Functionality.** DRS offers customers a subset of smart multiplexing capabilities that a small customer might not be able to otherwise afford – features such as drop and insert, rerouting on demand, and management reporting.

MCI will upgrade DRS to phase two in the third quarter of 1990 with a capability called "dynamic allocation of bandwidth". This will provide for management of shared facilities over fiber optic networks at the DS-0 or unframed DS-1 levels. Service will be provided over fiber optic trunks equipped with DS-3 automatic restoration and monitoring over most direct routes between endpoints. The capability is in testing now and will be tested on customers beginning in the second quarter, 1990.

MCI also provides another level of private line management through its MCIView. MCIView is MCI's gateway to IBM's NetView. It provides service and performance alarm information on MCI's TDS-1.5 service. Alarm information is displayed on the customer's IBM NetView Alerts-Dynamic screen. MCIView is another application of INMS – Operations Management.

Other Carriers Ready Products

Williams Telecommunications Group (WTG, and commonly known as Wiltel) is beta testing its own network management products. The company offers DACS service now, and will soon add reconfiguration options to respond to changes in network needs and disaster recovery situations. Wiltel is the first carrier to design its extended superframe (ESF) performance monitoring and reporting products to be proactive in nature – providing information to customers before they request it or before they know there is a problem. One new service that Wiltel is now testing is direct access via computer terminal for customers to its provisioning and inventory management systems. This will allow customers to check the status of any service order with the carrier, and compare progress relative to agreed-upon timetables. For instance, customers can see whether or not the access services have been ordered, confirmed, or installed, and if testing has begun. Wiltel will also allow customers to remotely enter service repair forms from the same terminal. The carrier is designing the system to alert users when they try to report a problem if Wiltel has already initiated corrective actions. In many cases, the carriers already know there is a problem when customers call in. Another bonus of this access is the ability for a user to review by his own computer the service repair history pertaining to their circuits.

Cable & Wireless will launch a series of network management functions in 1990, including CCR. Functions will closely correspond with the existing offerings in the marketplace, including standard reporting and graphical interfaces. The network management system will also allow users automated interface for trouble reporting, electronic mail and provisioning updates. This network management system will be integrated with existing fractional T-1 and T-1 products. Cable & Wireless is also testing its new "Bandwidth on Demand" service which will allocate switched bandwidth to customers as they require it.

User Pitfalls – Assumptions

As the carriers move to enhance functionality in their product lines, customers should be careful to examine all aspects of a certain function before adding it to their monthly bill. Too often, users overestimate capability when they are buying feature functionality from carriers. For instance, many carriers offer DS-1 facilities with Extended Superframe Format (ESF). This does not necessarily mean that the user would get ESF reporting from the carrier. Rather, the carrier will support ESF signals.

Furthermore, users should not assume the carrier is always the best source for a particular function. Take multiplexers for instance. Carrier multiplexing offerings just have not kept up with changes in technology. M24 or M44 multiplexing from a carrier will allow a user to divide and route 24 or 44 channels for a user – not much in today's bandwidth-intense world. A good multiplexer should be able to divide a T-1 into 96 voice or 100+ low-speed data channels. Customers should realize that the $200-$1000 spent monthly on carrier multiplexing might be better spent on more sophisticated equipment of their own.

DACS is another option that should be evaluated closely. While the benefits of using DACS for disaster recovery or automatic reconfiguration are clear, there is the question of who can provide these services better – the long distance carrier or the local exchange carrier. Several regional Bell holding companies have launched DACS services that can provide similar switching needs, but are geared towards the intra-LATA network. Users should determine the scope of the application first, then see which product is likely to better provide a solution.

Finally, users should watch route diversity and automatic alternate routing closely. Customers should not assume that if service is supplied through two different carriers that route diversity is assured. Many carriers buy capacity from each other and share the same circuits. Also, some carriers may provide route diversity by using another fiber pair, but in the same sheath. In this case, route diversity means little.

Take for instance a 1990 cable cut near Trenton, NJ. An old Lightnet cable, part owned by Sprint and part owned by Wiltel, was cut by a construction crew. The accidental cutting of cables was something that happens from time to time. However, in this case MCI, Sprint, Wiltel, Allnet and ITT customers were all using fiber that was routed together. One long distance reseller in the area had its routing tables set up as follows: Sprint, MCI, then ITT. All three carriers were having problems completing calls so the carrier was forced to send traffic over AT&T lines by equal access, an expensive alternative. Users should ask to see the physical routing of the circuit – it makes a difference.

New Providers Entering the Marketplace

Just when it appeared that consolidation was the trend in the industry, a few surprises have arisen lately. Newcomers have entered the private line services market, attempting to carve niches in the already crowded industry.

Southwestern Network Services (SNS) is trying to sell networking and equipment expertise with its core private line services. The carrier is trying to sell a total communications package, including network management and oversight, if customers want them. This is another example of a carrier trying to offer a non-generic product.

Another new (actually old) face is Southern Pacific Communications, Inc. (SPC), the company that originally started GTE Sprint, now US Sprint. The company recently completed a deal with MCI that traded right of way permission for a large amount of fiber capacity. With those circuits in hand, SPC is looking to re-enter the communications business by selling point to point private lines.

Other niche players are bound to continue to enter the market, driven by the temptingly low prices of T-3 circuits. It does not take a lot of expertise to maintain a private line resale organization. However, the emphasis will continue to be on added functionality and price. This price war cannot continue forever. Eventually some providers will collapse. One industry expert estimates that carriers charging rates 60 percent below AT&T are just barely covering short term costs and cannot remain viable in the long term.

However, the long term is quickly becoming shorter. Many private line providers are merely trying to stay in business long enough to be around when the Bell operating companies come hunting for national networks for their long distance transmission facilities. That could be as soon as 1994.

The Future of Private Lines

One indicator of the current standing of network management is the impostition by the carriers of network requirements on customers who wish to obtain true alternate routing or reconfiguration. Most carriers require the customer to have existing access and transport facilities in place, and rerouting is limited to switching circuits among the existing facilities. In the future, these features will be available on an "on-demand" basis.

Today, the interoffice carriers appear to be ahead of the BOCs in providing these services. For instance, look at a customer using MCI's DRS package. Suppose the customer had two sets of 56 kbps circuits, each running into an MCI DACS for termination in two separate regional data centers. The data centers are connected to the MCI network via a T-1. If one local access T-1 goes down, knocking out a data center, MCI could reroute those 56 kbps channels over the T-1 to the other data center. However, termination at that other data center would be limited to local access availability. If there is no room on the T-1 access channel, then the calls will be blocked at the second data center. Due to local access limitations the customer must prioritize applications to assure completion at a reasonable rate.

This is better than nothing, but to be effective in the future, rerouting must be flexible enough to provide service where no service exists now. When local access and interoffice transmission are available on an on-demand basis, in cases like the MCI example above, the calls could be terminated at the second data center through reconfigured access, likely using switched termination. This will not be easy for either interexchange or local exchange carriers to provide. Dynamic allocation of bandwidth – bandwidth on demand – is a problematic feature for both carriers. The provider must ensure that enough capacity is always available to serve the marketplace. In the initial stages, bandwidth on demand will likely be allocated on a "first come, first served" basis.

With increasing digitization of the local and long distance networks, transmission becomes more efficient, more error-free. ISDN based access will open a whole range of on-demand services. End to end spontaneous switching will become more a reality, especially with fiber to the premises connections. This is all in the future, however. For now, as users become more reliant on the digital backbone services – fractional T-1, T-1, and T-3 – network management and customer-controlled features will become more integrated on all levels. As fractional T-3 makes its entrance, it too will be logically extended to provide similar management opportunity. Carriers will outfit their T-3 DACS to provide T-1 on demand, in the same way that DS-0 can be provided on demand now.

The major carriers will offer more complete management across all private line products, not just for T-1. As these become available, the differentiation of service providers will become greater – helping to bury the concept of private lines as generic goods. In this case, resellers will find it hard to compete – they will be largely relegated to providing a pipe of capacity.

The biggest (and most obvious) outstanding issue for vendors of private line management is the same issue that their customers are grappling with these days – trying to get network management for all of the network managers. The optimal solution for any user is to get a network management system that can communicate, monitor, and control private line, virtual network, equipment, and other management systems. This is the big challenge for the 1990s, if not 2000s.

Chapter 8

Virtual Network Services

Virtual networks are rapidly becoming the standard long distance services for most businesses. They are the most sophisticated off the shelf products that the three major long distance carriers have to offer. Indeed, virtual networks open the door to a carrier's entire family of services in a single package – switched voice and data, private lines, travel cards, toll-free service, international service, and more.

Like 800 services, virtual private network services make use of innate network intelligence, stored software programs in carrier switches that allow users to create their own private telephone networks. In this sense, users can achieve a greater level of flexibility than with other hardware based services – an entire network can be reconfigured simply by changing a few parameters in the network database.

Virtual network services are quickly becoming the primary long distance service offered by carriers. While once reserved for elite high-end users, virtual networks are now being offered to companies with as little as $3,000 per month in long distance expenses.

On the surface, virtual networks can seem confusing, if not intimidating. There are all sorts of conflicting terminologies, complicated pricing schemes and chaotic feature plans. Recent additions of long-term pricing options have only served to further confuse the issue. However, the concept of virtual networks is not hard to grasp. As users come to understand the concept, they are finding that virtual network services can solve many of their diverse communications problems.

The Concept of Virtual Networks

Companies have long valued dedicated private line corporate networks for the prestige, control and technical prowess they symbolize, but the task of creating these extensive private line networks can range from the very simple to the extremely difficult. For a company with just two corporate locations, it is relatively easy: just calculate the economic break-even point between private lines and alternative switched long distance WATS services and the analysis is complete. If the cost of switched WATS service between the two sites exceeds the cost of private line, install the private line.

For multisite companies, however, networking can be an overwhelming task. Telecom managers have to decide which sites have enough volume to warrant high capacity WATS lines, which ones should use regular switched long distance service and which ones should be linked by private lines. Users must then evaluate the various multi-site pricing plans available from the different carriers, tying in the various long-term and discount options. Additional factors such as time and capital budget constraints, not to mention the regional availability of carrier services, can further complicate the problem. Resulting multisite networks typically involve different carriers, services and forms of access. Monthly bills can be confusing. Transmission failures can be a nightmare, with the different carriers pointing fingers at each other.

Today, having a virtual network is just as prestigious as having a vast private line network, and probably more economical too. Users are relieved of major staffing and maintenance requirements without losing control over their network. Customers get on-line access to the carrier computer network databases so they can reconfigure their networks as needed. Virtual networks also allow a company's smallest sites to become tied into the corporate network. In the past, these small locations have been kept off the private network because of the high cost of dedicated circuits. With virtual network services, other forms of access, namely switched access, allow these small sites to join larger locations in forming a cohesive communications network through a single carrier.

Perhaps the best characteristic of virtual networks is that they are now available for companies with moderate monthly bills, i.e., $10,000 or more. In some instances, especially government networks, virtual networks may be installed in cases where monthly costs are as low as $3,000. While the first virtual network services were targeted towards private line users who wished to augment or entirely replace their

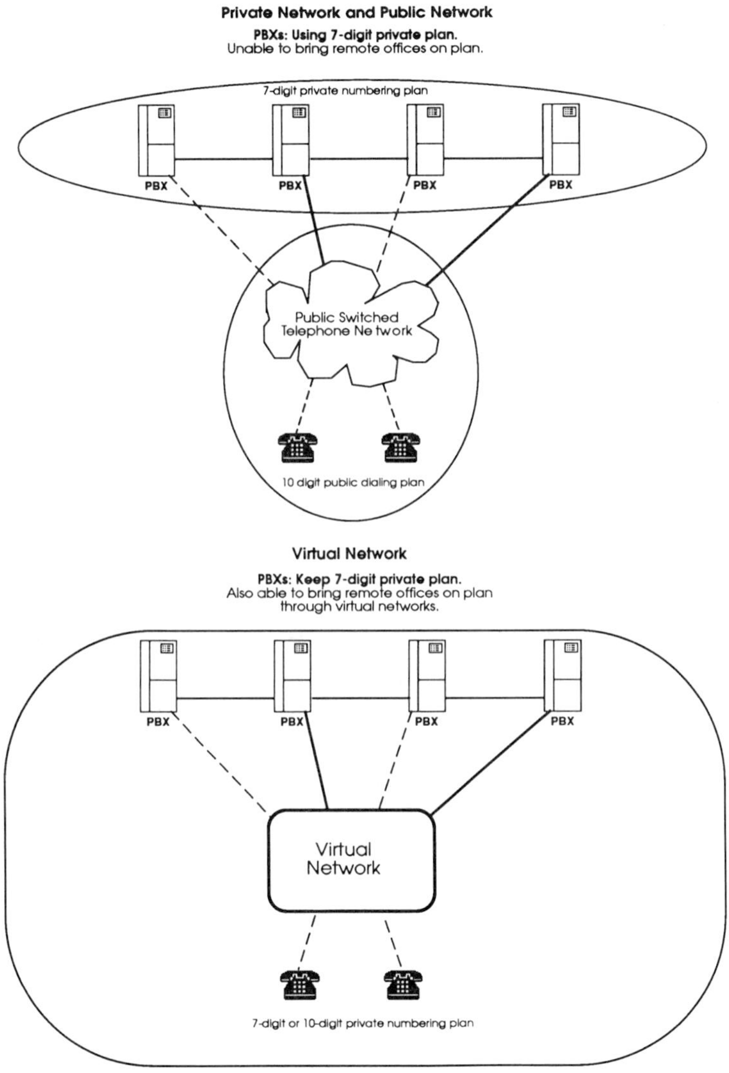

Figure 8.1 Comparing private networks with virtual networks.

private line networks, these services have grown to attract smaller multisite switched WATS users as well.

Network Architectures

The sophistication of the various virtual network services varies according to the degree of intelligence in each carrier's network. The ability to provide advanced routing capabilities and enhanced screening features, for instance, is dictated by the degree to which each carrier has developed a strong foundation of supporting software and signaling protocols to handle those features.

The signaling system that each carrier uses represents a large part of this intelligence. Each of the carriers is migrating to or has installed signaling system 7 in its network to support advanced forms of routing and processing of calls. Particularly with regards to virtual network services, signaling system 7 is a necessary precursor for installing an integrated services digital network capabilities in a network. US Sprint and MCI have completely installed SS7 system-wide; AT&T is steadily upgrading its network and should be finished by the end of 1990.

As far as virtual networks are concerned, faster call set-up times and better network utilization are the most immediate benefits of SS7. Call set-up time represents the time it takes for the network to look at a call and determine the proper routing for that call. All of the carriers deploying SS7 have seen decreases in call set-up time over previous signaling methods, although AT&T, with its Software Defined Network (SDN) service, has seen less of an improvement since it was already using a fast signaling protocol, signaling system 6 (SS6). US Sprint, with its Virtual Private Network (VPN) product, reported decreases of more than two seconds after its migration, and MCI claims it saw similar reductions with its service, Vnet. Nevertheless, AT&T still has the fastest call set-up time in the industry, averaging between 4.5 and 5 seconds per call – beating MCI and Sprint by nearly two seconds each – because AT&T has direct connections to more end offices than the competition. As these connections become available to MCI and Sprint, the two carriers will buy them, further reducing the difference in call set-up times.

Different Network Priorities

Signaling system 7 is one part of many overall network upgrades that are underway by each of the carriers. Each carrier network is in a different stage of development, partly due to differing priorities set by upper management. In general, AT&T and

Different Types of Virtual Network Calls

In the past, sites that had dedicated access were generally referred to as "on-net" sites, and all other access was "off-net." However, with the advent of equal access into the virtual networks, these sites fall into a grey area – they can be on-net from a corporate dialing standpoint, but may be charged higher off-net rates.

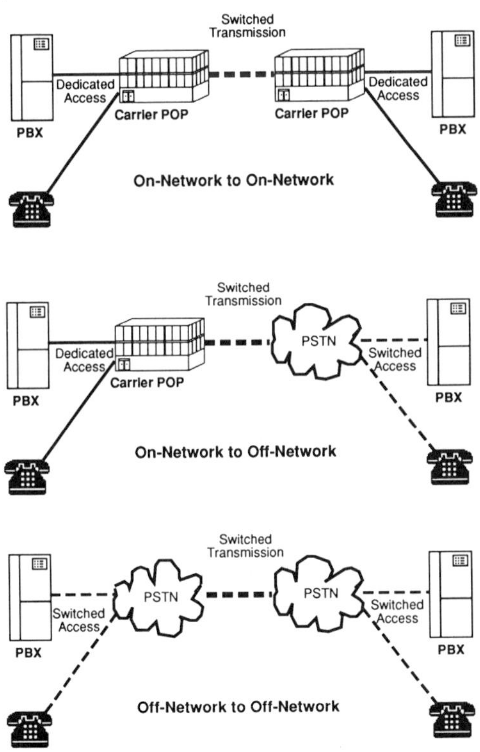

Figure 8.2 Different types of virtual network calls.

MCI have sought to expand upon network intelligence while US Sprint has sought to develop its physical network.

How these different priorities translate into differentiated product offerings is exemplified in the way each carrier announced switched 56 kbps service. For a long time, AT&T offered switched 56 service through its Accunet Switched 56 product, but at rates that were considered by most to be extremely expensive. This was due to the fact that AT&T had to carry the calls over a separate all-digital network. When US Sprint announced its switched 56 option, it announced rates nearly 90 percent below those of AT&T. This was possible because Sprint sends these data calls over its standard voice network which is 100 percent digital. When MCI offered its service, it too had to send calls over a separate digital data network. MCI, however, is meeting the competitive pricing of Sprint despite the separate network. Both MCI and AT&T are migrating to a single all-digital network.

Carrier Network Capabilities

As a result of these different directions in network development, there are notable distinctions among the virtual network services, particularly in the area of enhanced processing of calls.

Dialing Schemes

One of the more basic features among the network offerings is the available dialing schemes for each service. A dialing plan is a company-wide definition of what phone numbers will look like. Are all calls going to be ten digits in length, seven digits, a combination of the two?

Seven-digit plans are typically associated with firms that have had a private network in place for many years. Seven-digit plans allow a company to differentiate between traffic traveling over the company network ("on-net") and traffic traveling off the company network ("off-net"). Corporations traditionally translated ten-digit public switched telephone network numbers into seven-digit numbers for all those destinations on the company network. Among other things, this shortened the dialing requirements to reach some of the most frequently called numbers.

Ten-digit plans are the standard for most public networks. It's the format now used by all companies who are using the public switched telephone network to complete calls. A corporate directory using a ten-digit plan is nothing more than a listing of all employees' standard telephone numbers. Only a few companies with private networks maintain ten-digit numbering plans that are different from those in the public network – there are few benefits to this.

The beauty of virtual networks is that they can handle a number of different dialing plans, depending on the needs of the corporation. Network software allows users to store corporate dialing directories in databases for translation. This enables:

- Seven-digit private networks to retain their present directories, and even bring the small outlying sites onto corporate networks, either as ten- or seven-digit dialers.
- Ten-digit users to keep dialing ten-digit numbers, as they always have.
- Telecom managers to have mixed corporate dialing plans. This helps them to cope with personnel who insist, for whatever reason, on dialing ten digits even though there is a seven-digit internal numbering plan.

Anytown Department Store Network Dialing Plan

Store:		Departments:	
East End:	346	Main Operator:	0000
West End:	741	Fax Machine:	4FAX
North End:	441	Personnel:	HIRE
South End:	829	Accounting:	CHEK
		Computer Room:	DISK
		Courtesy Desk:	HELP
		Sporting Goods:	SPRT
		Electronics:	ELEC
		Automotive:	AUTO
		Men's Wear:	MENS

For example, to call the Courtesy Desk at the North End Store a network user would dial, 441-HELP (4357).

Figure 8.3 An example of a customized seven-digit network dialing plan.

A dialing plan can be a creative source for solving telecommunications problems. For instance, look at a retail chain that identifies its stores by three-digit numbers. The virtual network could be set up with a seven-digit plan so that the exchange corresponds to the location code and the station number spells a department name. Employees could reach any location's main operator number by dialing a three-digit location code plus 0000, or any site's facsimile machine by dialing the location code plus 4FAX.

Depending on the situation, seven- or ten-digit plans could be appropriate. Companies typically look to maintain consistency with existing plans in order to ease transition to the new virtual network. Presently each of the carriers offers a full slate of ten-digit and seven-digit on-net dialing plans, as well as ten-digit off-net dialing plans, for customers.

Conversion Capabilities

An important aspect of virtual network functionality is conversion capability, that is, the ability of the network to translate one number into another number. In some instances, conversion is critical in providing advanced dialing functions, such as the mixed dialing plans discussed above. In other cases, as with forced on-net routing, specific conversion capabilities like ten-digit to seven-digit translation is required in the network to even offer the function.

Each of the carriers can perform seven-digit to ten-digit translation, as well as ten-to-seven, seven-to-seven, ten-to-IDDD, and seven-to-IDDD (IDDD stands for international direct distance dialing).

Access Methods

Another big advantage of virtual networks is their ability to serve all types of calling needs with a variety of access methods. After all, not every location can afford to install a dedicated private line. For small, out of the way sites, access can be a key factor determining which carrier to choose. Dedicated access is fairly standard among the three carrier services. For the most part, all three carriers offer similar dedicated access services. The true differences come in the switched access and travel card access into the networks.

AT&T and MCI use their respective equal access codes to enter a virtual network via switched access. AT&T uses its secondary equal access code, 10732 instead of 10288 (10ATT), because it needs to separate those calls that are destined for high-end business services, such as SDN and its Hospitality Network Service (HNS), from the standard residential and small business 10288 calls. MCI uses the same 10222 equal access code for all of its services because its switches can screen the calls as they enter the switch. Most of the time, users do not have to dial these 10XXX access codes – the local exchange carriers assign those codes automatically to each telephone line presubscribed to a particular carrier, or most PBXs can automatically insert the digits in front of dialed digits.

AT&T and MCI users have the option of dialing a seven- or a ten-digit telephone number through this form of access. Dialing a seven-digit number requires users to precede calls with a 1 plus 700 special access code which tells the local telco to route the call to the presubscribed line's long distance carrier, in this case either AT&T or MCI. Users dial a ten-digit number in the standard 1 plus number manner.

US Sprint's feature allows users to dial a "1" plus a ten-digit number to complete a switched access call, like AT&T and MCI. Customer locations will be signed up as "Dial 1 WATS" locations (Dial 1 WATS is Sprint's equal access WATS product). To complete calls to on-net points, Sprint will perform a ten-to-seven translation and route the call accordingly. Sprint does not plan to offer a 700 form of access.

For remote or travel card access, each of the carriers has a slightly more differentiated range of offerings. AT&T gives each SDN its own 800 number that users dial to access the network. Users are prompted for personal ID codes before gaining access to the network to place seven- or ten-digit, as well as IDDD, calls. Also, AT&T has an "express" option that terminates an 800 number directly at a specific on-net station, without requiring an ID code entry.

MCI offers both Vnet Card and a remote access option. A user dials a customer-specific MCI Vnet 800 number and places a seven-digit or IDDD call. As with AT&T, users are required to enter an ID code. Importantly, Vnet Card users can be restricted to certain calling privileges through MCI's customized range privileges option.

US Sprint offers its standard FONCard option, in which a user dials Sprint's 1-800-887-8000 number and inputs an ID code plus a ten-digit number. Sprint just announced a FONCard that accepts seven-digit numbering plans – a so-called "On-net only VPN FONCard" that restricts callers to only VPN on-net points, reducing the potential for fraud.

Carrier Feature Options

Many telecom managers place a great deal of emphasis on the functionality provided by the virtual network services – the ability to block calls from this place to that; to track calling from a particular ID code; to access the network in a particular way, etc. In fact, while virtual network pricing will become extremely competitive over the next few years, feature competition is likely to become just as fierce – largely due to the demand in the marketplace.

Realize that a carrier's ability to offer a particular feature is determined by the level of software intelligence in the network's switches. Often, alterations to one portion of the software can yield a multiplicity of changes to a virtual network service in general. When US Sprint, for instance, finalized its "extended digit screening" capability that allowed it to screen the full ten digits of any call in its virtual network, the carrier was able to introduce ten-digit on-net dialing plans, ten-to-seven conversion, switched access, forced on-net routing, and other similar features.

Here is a sampling of the feature functionality presently available from the carriers (note that not all carriers have all of these features):

- **Account codes.** These are the same basic two-, three-digit, etc. codes used with WATS services to allocate costs among cost centers. Typically, a user dials the 1+telephone number and then receives a second dial tone prompting for entry of additional digits. At the end of the month, the bills are sorted according to these supplemental digits.

- **Authorization or ID codes.** These are similar to account codes in that they are generally entered after the dialed telephone number. However, these codes are validated – i.e., checked against a user-defined database for calling authorization or restrictions. The call is only initiated by the switch upon entry of a valid authorization code.

- **Call screening.** Each of the three carriers offers some sort of account code and authorization code feature. What is most significant is the ability to assign calling restrictions based on these codes, or even based on trunk numbers, automatic number identification (ANI), etc. These are commonly referred to as "class of service restrictions", or "call screening".

  If a calling card is lost or stolen, for instance, a telecom manager can go into his virtual network database and assign a "no calls allowed" restriction to the card, turning off service automatically. Or if an employee moves to a different department where he will have broader calling needs, his card can be adjusted to allow calling to wider ranges of coverage.

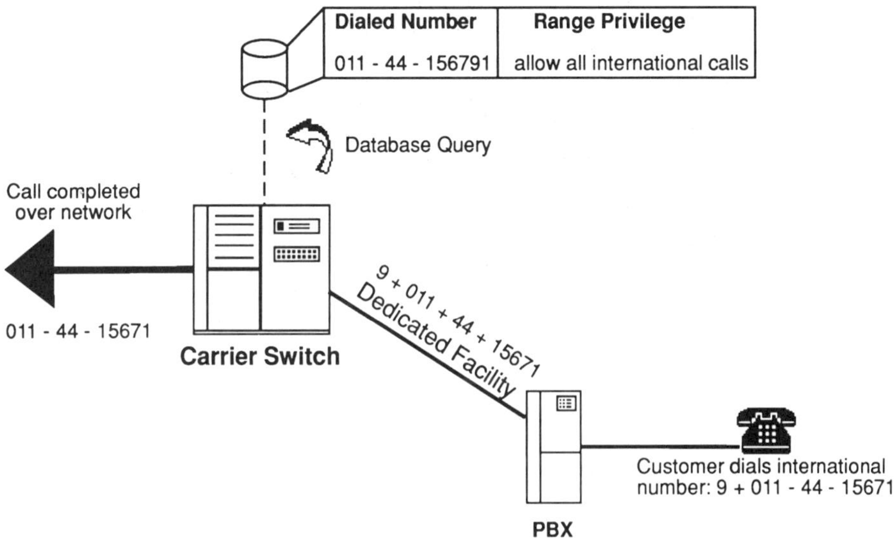

Figure 8.4 A typical call screening arrangement.

- **Public switched network overflow.** With the public switched network overflow feature, carriers can convert a seven-digit on-net number to an off-net ten-digit number when the on-net line is busy or out of service. Each of the carriers offer this feature; however, execution varies among the carriers.

  Suppose a user dialed 555-1234. The network screens all seven digits of the call, searches for the translation in a table, and completes the call with a different exchange and NPA. So, the call might go to 804-345-1234 after this process is complete.

- **Forced on-net routing.** This feature allows a company to retain its ten-digit calling plans for on-net calling. Basically, the network translates the ten-digit numbers to valid seven-digit numbers for on-net comple-

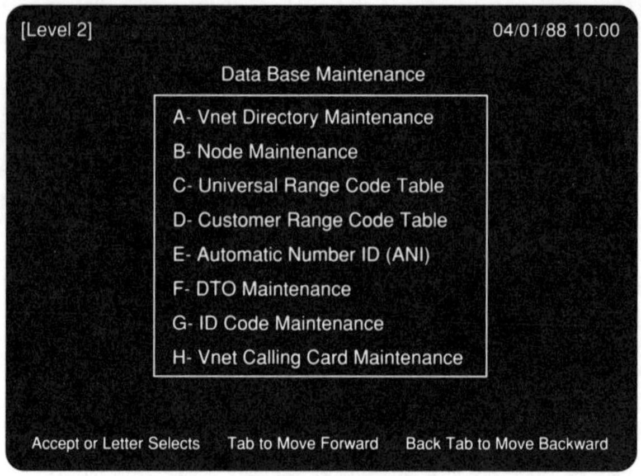

- **View, activate, deactivate and modify Vnet numbers**
- **Enter changes with a scheduled effective date: immediate or deferred**
- **Access via VT-100 terminal or personal computer with emulation software**

Figure 8.5 Looking at MCI's network management package — the main menu (from Briere, D., Virtual Networks, © 1990, Artech House, Inc.).

tion. This gives the telecom manager a greater level of flexibility in organizing his dialing plan.

- **Internetwork dialing.** A variety of features from each of the long distance carriers allows users to interconnect the virtual network with other virtual or private networks. This feature is vital for the survival of any corporate telecommunications network in an era so marked by mergers and acquisitions. Simply, these allow the partitioning of networks into operational sub-networks while still maintaining the cost efficiencies of communicating as an integrated network. When one division is sold, that sub-network is merely separated from the overall network, and each unit maintains its communications plan and structure.

- **Network intercept announcements.** These enable the customer to have calls intercepted by prerecorded virtual network specific announcements to relate calling restrictions or network conditions. These announcement can bea standard announcement by the telephone company or a customer specific announcement. A company might have a recording such as "I'm sorry, but your call for the XYZ Corp. cannot be completed because of calling restrictions on your personal calling card."

- **Network remote access calling.** This feature allows users to call on-net or off-net sites from off-net points, typically by dialing an 800-number, punching in the target telephone number and authorization codes or ID codes, and then placing the call.

- **Network management systems.** These are critical in trying to migrate private networks to virtual networks. Telecom managers are used to having a great deal of control in a private network – control they must keep if moving to a public network offering. The sophistication and depth of these network management tools is quite extreme and getting more so each year.

- **Speed dialing.** US Sprint is the only carrier that has adapted this WATS feature for the virtual network market. If users dial certain numbers more than others, the telecom manager could program his VPN so that these numbers could be converted to short three-digit codes. Neither MCI nor AT&T has this capability today.

- **Switched data services.** Switched data is an all-important feature for virtual networks. Without this functionality, many private networks would stay private. Virtual network services are still developing and enhancing this capability. Switched 56 kbps service is the first foray into switched data services and is a rather new item in the integrated virtual network context. Each of the carriers announced extended development programs for switched data in mid-1990, including switched 1.536 Mbps transmission by the end of 1991.

  AT&T announced in 1988 that it could link SDN with its Accunet Switched 56 kbps network, and both MCI and US Sprint recently announced the switched 56 feature for their networks as well. But Sprint is the only one of the carriers that is presently sending the switched 56 traffic over its voice network as standard voice calls. AT&T and MCI send data over digital sub-networks as data calls. MCI has been quick to allege that the Sprint method degrades the product, but in many

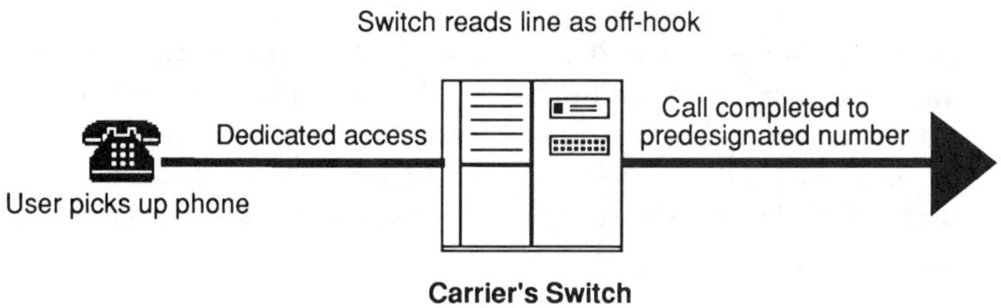

Figure 8.6 Virtual network ringdown circuits are found in most airports.

ways, it is a sign of the times to come – a complete marriage of voice and data over the same digital facilities. Both MCI's and AT&T's networks are evolving to the point where they will send mixed voice and data over the same facilities, like Sprint.

- **Virtual network ringdown circuits.** This feature allows a user to have a telephone directly connected to a carrier point of presence so that when someone picks up the telephone, the switch will automatically dial a specified telephone number. This number may be on-net, off-net, or even international. The application is for companies such as airlines or financial institutions that wish for an immediate connection to a certain number. Most airline and hotel phones in airports work this way.

- **Virtual foreign exchange (FX) circuits.** These allow companies to maintain a "sales presence" in a distant city. Callers in that area dial a local number, which then connects with the company by way of a long distance virtual network connection. It works the other way too – a company in Miami could maintain a virtual FX to New York because of the number of calls to the metro New York calling area. This would cut down on costs between the two areas.

The Bottom Line

What carrier is best for a particular user is tied directly to that user's needs, not to which carrier may offer the best overall virtual network service. Each carrier has strengths and weaknesses and the mindful telecom manager will match his needs with the carrier that best meets those needs. Too many managers are tempted by all the extra features and options provided by the different virtual network services, when in reality, few of these options will ever be used.

While AT&T and MCI definitely have a broader menu of features, most of these are not be needed by the average company, bringing transmission quality and price more into the picture. In these cases, US Sprint might be a more attractive alternative. However, you may find your company has such intricate routing requirements that only one of the carriers is even in the position to meet them.

Be aware that each of the carriers has a wealth of talent to produce customer-specific features within your network. This is the beauty of a software-defined network. Many features are provided off-tariff or on contract. Users should make it a point to tell carriers their "wish list" as far as feature functionality is concerned. In many cases, the carrier can adjust its system to meet the customer's needs.

Ultimately, users should determine what their current and future requirements are likely to be, then concentrate on how each of the carriers can respond to their needs. Without this, users could spend a lot of time trying to determine who is the "number one" virtual network provider in the market and drive themselves crazy.

The Future of Virtual Networks

The virtual network market will continue to address lower segments of the long distance marketplace. There will be pressure on the long distance carriers to repackage the virtual network services to meet niche market needs, however. A good example of this is the MCI Vision product, launched in the summer of 1990. This is effectively a mini-Vnet product, offering some of the basic Vnet features with modified pricing and billing. It is designed to attract small and medium sized business users.

More emphasis will be placed on service and customization as well – all part of the overall market move away from price competition to service competition.

Chapter 9

Calling Cards

Calling cards have been familiar to phone users for some time. They are called by many different names – calling card, travel cards, credit cards, etc. – but they all share the same function. They allow users to make calls when on the road and charge them automatically to their home or business account.

The AT&T Calling Card, originally known as the AT&T Credit Card, the template of these services, was introduced in the early 1970s. It is one of the older long distance products, far eclipsing in age many of the newer services such as mileage-sensitive business line based WATS products, or switched access-based 800 services.

Many analysts have been quick to label the calling card market a mature one. In some ways this is true – the percentage growth changes from year to year are not as dramatic as with some other products. However, the calling card market is not standing still. Recently there have been some radical changes in this market, changes that the carriers hope will make travel card calling more simple, convenient and functional to the caller. At the same time, the new trends create a number of pitfalls that card users should avoid in order to make travel card calling both cost effective and hassle free.

The Products

The calling card market is both wide and varied. There are many possible suppliers of essentially the same service – interexchange carriers, Bell operating companies, independent local exchange carriers, aggregators, and other third party service bureaus. Many of these companies also have multiple products.

Travel cards may be marketed as either stand-alone products or as part of another service. They also can have their own discount schedules for volume calling or their rates may be included as a part of a related long distance offering, such as WATS or virtual network services. Look at the variety of options for calling card users:

- **Stand alone products.** These are cards that have their own account separate from any installed line accounts. Indeed, a customer does not have to be a subscriber to any other long distance service in order to have a calling card under these programs.

- **WATS-Based Products.** These are cards that are directly integrated with carrier WATS programs. As such, long distance usage through travel cards is usually counts towards meeting WATS volume discount levels – meaning higher discounts for all calling under the WATS plan. With other WATS based card products, however, a different schedule of discounts may apply to the calling card traffic portion of the bill.

- **Virtual network based products.** These represent the top of the line in calling card services. Virtual network cards offer users direct access to the various functions of virtual network services from a remote location. Features such as call screening and forced on-network calling allow the telecom manager to control how calling cards are used. Managers can assigned certain calling restrictions to each card – on-net only, domestic only, international, etc. Users are also offered a number of different access options. Usage from virtual network cards are typically included with the virtual network discount plans, resulting in substantial discount levels for card calling. The AT&T Calling Card, MCI Card, and Sprint's FONCard all have their virtual network counterparts.

- **Fax network based products.** Similar functionality is found with facsimile network cards. These cards allow a user to access and use facsimile network offerings from the road. Once users have accessed the facsimile network with a travel card, they can perform most of the same facsimile network tasks as when in the office. From remote locations, users can send and receive facsimile messages, change routing lists, and request report information. An example is the MCI Fax Card.

- **Credit calling cards.** Hailed as the cards of the future, these multi-purpose cards offer users both telephone calling and banking privileges. These cards are marketed as a "one-stop" solution for users. Users could conceivably trade in their VISA card, automated teller machine (ATM) card, and telephone calling card for a single multi-use credit card.

The carriers are all trying to be flexible in how they position their card products for each segment of the market. The market trend is to integrate these different functions into one card so that a user's calling card is truly an access mechanism to the carrier's entire line of long distance services. There is no need for separate facsimile network cards, virtual network cards, and stand-alone calling cards – they all access a calling card database for handling instructions. In the future, more than now, these different services and capabilities will be integrated into one card, although different access numbers might be used to access different services.

The History of Calling Cards

The different calling card products available on the market today are at different stages of development within each of the carriers' product lines. Modern competitive calling cards, i.e., those offered by all of AT&T's major long distance providers, trace their roots back to the days before equal access was universally installed around the United States.

AT&T created the product. The AT&T Credit Card was created to take some of the pressure off of company operators in handling home-number billed calls. The Credit Card took advantage of changes in technology that let the user input the numbers via touch tone telephone keypads.

The initial AT&T Credit Card operated in much the same manner as today's. The user dialed 0 plus the telephone number and entered the calling card number when prompted. This was the standard against which all other travel cards were evaluated. AT&T used a proprietary database to validate the calling card authorization codes and to process calls. All calling card numbers should be validated before a call is placed to avoid fraud. Fraud is a major concern within the industry as millions of dollars each year are lost when stolen authorization numbers are used to make unauthorized calls.

Company	Product	Type
AT&T	AT&T Card	Stand alone, WATS-Based, virtual network based
	AT&T Universal Card	affinity card
MCI	MCI Card	Stand alone, WATS-Based, virtual network based
	MCI Fax Card	fax network based
Sprint	FONcard	Stand alone, WATS-Based, virtual network based
	Sprint VISA	affinity card

Figure 9.1 Available calling card products from major long distance carriers.

In the first days of long distance competition, AT&T's service charge was $1.15 per call for using the card. This covered the additional expenses of maintaining a database, card production, and other service-specific costs.

The AT&T product caught on quickly. The AT&T Credit Card became universally accepted – everyone learned how to use one and AT&T used its all-important BOC and other local exchange carrier relationships to make universal access and billing a reality.

Over time AT&T changed the name of the card from the AT&T Credit Card to the AT&T Calling Card. The carrier began issuing plastic cards with a magnetic reading stripe on the back instead of paper, tying the product with its own AT&T Card Reader phones now seen in airports, hotels, and other locations. By encouraging customers not to punch in their AT&T Calling Card numbers, the carrier was potentially cutting back on fraud as authorization code pirates could not actually see a person entering the digits. AT&T steadily lowered its service charges as improvements in database technologies and operations lowered the costs of providing the validation function, as well as for sheer competitive reasons. As of August 1, 1990, this service charges was $0.80 per call.

AT&T's Competitors

It is against this backdrop that AT&T's competitors started to offer their own travel card services. They were handicapped from the start. Before equal access was universally available, customers of alternative long distance services – GTE Sprint, MCI, SBS, etc. – had to dial an access number plus a host of additional account and authorization numbers to access that carrier's network and place a telephone call. While autodialers may have helped at the home or office, they were of little help when on the road. A market quickly developed for small hand-held tone generators that could be programmed to remember the string of local access and identification digits.

The first methods of accessing carrier networks by card mimicked local access options for homes and businesses. At the time, feature group B circuits (950-0XXX or 950-1XXX) were used for local access. The XXX corresponded to the carrier identification number issued by BellCore so they were nationwide numbers. Thus, an MCI traveler could dial 950-1022, enter telephone number and authorization code, just as he or she did back home. In most cases, when providing access in this manner, the carriers were not adding much overhead to the switch maintenance function, so there were no surcharges added.

This was a big advantage for the alternative carriers and their customers. If users did not mind the extra digits, they could save a substantial sum in surcharges. In many cases, the tone generators eliminated the digit nuisance. The only drawback was that the traveller had to be in a city where the carrier had facilities.

In time, the carriers converted to longer-term travel card networking options, revamping the services to bring them in line with the changing local and long distance network environment, as well as to make them more profitable as stand-alone services. The carriers began to surcharge for travel card calls and, in some cases, charged rates above their regular direct distance dialed (DDD) rates. Generally, however, the cost of a call was still below AT&T.

Access is still somewhat limited. As will be discussed later in this chapter, "0 plus" dialing is available on a limited basis for alternative carriers – whenever the phone used is presubscribed to that carrier network. Otherwise 950 and 800 access are still common. Users dial those numbers, receive a network tone, dial the telephone number to be called, receive a second network tone, and then input the authorization number.

Each of the competitors have started offering a host of innovative features for their card products. One major way is to offer vanity cards – plastic cards that have the customer logo and design imprinted on the card, instead of the logo of MCI, Sprint or whatever the carrier might be.

There are other unique products being developed for travel cards. Consider the following:

- US Sprint has a Braille version of its FONCard, offering more accessible long distance telephone service to the blind. All numbers necessary to complete a call are encoded on the card in Braille. There is no extra charge for the Braille FONCard, which also features special services including free directory assistance and a free over-the-phone bill review.

- Sprint has also announced a voice-activated calling card, which bypasses dialed digits in favor of voice commands. A user could say, "This is Mark Langner. I wish to call home." The network would verify Mark's voice against a digital record, then look up the number for "home" and place the call.

- Telecom*USA's calling card allows users direct access to enhanced services such as voice mail and conference calling services. If a person cannot be reached, the caller can activate a voice mail system to store a message for automatic transmission later to the number called.

- MCI has a special calling card for its MCI Fax offering that truly gives users the same functionality on the road as at the home office. This is critical for sending facsimile messages from personal mailboxes to a facsimile machine at a hotel or other remote office. With additional authorization and security code features, the MCI cards can provide an important safety buffer to make sure unauthorized persons do not read facsimile messages.

These are just a few of the ways that AT&T's competition is seeking to distance itself from the market leader. As enhanced services become more important to the

Figure 9.2 Sprint is differentiating its product with unique features. The carrier has introduced a Braille calling card as well as a voice-activated card.

carrier product lines, travel cards will provide a significant method of remote access to those services.

Bank Cards and Calling Cards

In the face of this heavy competition, AT&T is fighting back. In 1990, AT&T launched what appears to be one of the hottest topic in travel cards in 1990 and 1991 – the AT&T Universal Card. The Universal Card is a combined VISA or MasterCard and AT&T calling card. The new cards bear the AT&T name and corporate logo but are issued by the Universal Bank of Columbus, GA.

The AT&T Universal Card is an affinity card; that is, it carries the name of a company or organization that helps promote its use in return for a share of the profits earned by the issuing bank. Banks issue similar cards for airlines, university alumni associations, athletic teams, and other charitable groups.

Sprint also offers an affinity card, called the Sprint VISA card, through State Street Bank and Trust Company, Boston. Quietly introduced in late 1989, the Sprint VISA card is a combined VISA card and FONCard. Customers can pay their FONCard bills with the VISA portion of the card. Otherwise, the Sprint VISA card acts just like a standard credit card, with features such as accident insurance, rental car discounts, emergency airline tickets, etc.

MCI launched its Card Compatibility program in July 1989, with the introduction of VISAPhone. This program allow VISA cardholders to use their existing VISA card to make long distance calls over the MCI network with billing to their VISA card account. The VISA Account number acts as an MCI calling card number.

This differs from the affinity card offerings by AT&T and US Sprint. The affinity card offerings by US Sprint and AT&T compete against existing bank affinity programs. MCI's Card Compatability allows multiple banks to enhance their card offerings by including telephone calling.

Carriers are hoping to add additional revenues through these affinity agreements. However, it can raise problems too. Carrier customers such as American Express and other bank card issuers now see these long distance vendors as their competitors – and are switching their traffic to competing carriers without affinity programs.

BOC and Independent Cards

The long distance carriers do not offer the only cards in the market. The Bell operating companies and independent telephone companies also offer their own version of calling cards. Many of these cards have been around as long as their IXC counterparts. The local telephone company cards offer many of the same functions as the IXC cards. The cards are ideal for customers who may not have a IXC card or who may not be on a program that has one. Additionally some BOC cards are accepted by multiple IXCs giving users additional flexibility as to where they can call from.

The local exchange companies are somewhat hindered in this market in the same way they are hindered in offering general long distance telephone services – they cannot carry inter-LATA calls. These must be carried by a presubscribed long distance carrier. In cases in which an inter-LATA call is placed, the local telephone company must transfer the call to the presubscribed inter-LATA carrier for completion of the call.

As a result, the local exchange companies are accentuating enhanced functionality of their cards. For instance, Bell Atlantic and BellSouth currently are working to develop "smart cards" which use special pay phones and feature:

- **Alternate account billing** – allows a customer to bill a call to one of two accounts already programmed into the card.

- **Personal phone book services** – stores 50 phone numbers in a microprocessor, embedded in the card, and allows speed dialing.

- **Office dialing** – stores names and numbers of four people and allows speed dialing within an exchange.

Early tests of these microprocessor based cards have been successful, leading some industry gurus to speculate that microprocessors will replace magnetic strips on calling cards in the near future. The microprocessors interact with software and hardware in the pay phone equipment, just as magnetic tape strips work with special magnetic strip reading phones.

Pay phone Presubscription

The development of calling cards and their relationship to the entire equal access process is tied to developments in the regulation of the pay telephone industry. A vast proportion of calling card calls are made from pay phones and there was considerable concern on behalf of AT&T's competitors about the bias of the pay phone process towards AT&T.

Until 1989, all interexchange "0 plus" calls from a pay phone were sent to AT&T, simply because it had always been done that way. For users to access MCI, Sprint, or another carrier, they had to dial a 950- or 800-number. In late 1988, Judge Greene ordered the BOCs to stop favoring AT&T. He asked for the opinions of the various industry participants as to how the process could be equalized. After much discussion, Greene established an interim process in which all public pay phones would be presubscribed to a selected carrier – all interexchange calls from those phones would be processed by the presubscribed carrier. As to who would make the long distance carrier decision, Greene ruled that the proprietor of the premises where the public phone was located would decide which carrier should be subscribed to the pay phone. Privately owned pay phones were not subject to Judge Greene's order.

The recent ruling for equal access from pay phones has helped little to alleviate this bias except in terms of dialed digits. The only positive effect for AT&T's competitors has been that, in many cases now, AT&T customers must dial 10288 before placing a call from a phone that is not presubscribed to AT&T. This equalizes the number of digits that customers have to dial. The downside is that AT&T won 80 percent of the presubscribed pay phones, so odds are that the customer is just dialing 0 plus the number anyway.

The Key Role of Signaling

Once the call travels to a particular network, calling card numbers must be validated. Validation is as vital an issue to travel cards as number portability is to 800 services. To validate calls, the local exchange companies interconnect with validation firms such that database queries travel from the local network to validation firm's switch. As with other services requiring database access, network signaling plays an important role with calling cards.

Signaling system 7 is used by the long distance carriers to compare validated numbers with those contained in a master customer database. Local exchange carriers are using similar signaling systems to access their respective databases. It is expensive for each local carrier to establish its own validation databases so most independent telephone companies use several industry BOC and non-BOC validation firms for validation services.

The BOCs are awaiting the development of the all-encompassing national Line Information Database (LIDB), discussed in Chapter 6 on 800 services. Each of the BOCs has been developing its own LIDB database for interconnection in 1991 or 1992. These interconnected databases will allow any BOC to validate the card of any other BOC; likewise, the interexchange carriers will have access to the BOC numbers for validation as well.

Nearly 300 independent telephone companies are using an SS7 network run by US Intelco to give their cards much of the same functionality as IXC and BOC cards. These independent carriers are placing the numbers of their calling card subscribers into the US Intelco database for storage and access. The US Intelco card, called the TelCard, displays the name of the telephone company along with the TelCard logo.

Problems with Validation

The pay phone presubscription process approved by Greene may help alleviate the problems of all "0 plus" calls being carried by AT&T. However, it does not help AT&T's competitors to promote their own calling cards in the place of the AT&T Calling Card. The reason lies again in validation. There is no system in place by which the BOCs can determine the carrier which issued a card and route the call over the appropriate carrier network. Different carrier numbering schemes exist.

The BOCs are well established to validate AT&T card numbers, since they use a similar numbering system. The formats used by AT&T's competitors do not follow the same format; rather, they conform to their own in-house proprietary formats. The practical result of this is that it is a relatively simple task for a competitor to accept and bill for calls using the AT&T or BOC numbering scheme – the billing number is in the calling card number itself. For the same reason, it is not so easy with the other formats – there is no way for MCI to trace the owner of a Sprint number. Thus, this makes it far easier for a user to make calls using cards using the AT&T or BOC numbering format.

The BOCs have said that if AT&T's competitors wish to convert their numbers to the BOC and AT&T format, then they will store them in their databases. However, this blurs the distinction between the cards of the various carriers. The carriers do not only want to have their cards validated, but also want to carry the traffic – this is how they make their money. Since each IXC's card would be accepted on any competitor's interexchange network, it would defeat a main reason for having carrier-specific numbers, to encourage greater traffic over a particular carrier's network.

Another suggestion is that all of the carriers – AT&T, MCI, Sprint, etc. – change to a new numbering system. All calling card numbers would be unique and each number could be associated with a particular carrier. However, there are nearly 100 million cards in circulation and replacing all of them would cost an enormous amount of money. In the event that a universal system could be adopted in the

U.S., there is an international 19-digit calling card standard to which the carriers could convert; however, this requires five digits more than presently used.

No one can provide an easy solution to the problems of calling card validation. When the LIDB databases come on line (they have been delayed several times), some of the problems involved with calling card validation will be solved. Yet, with each problem solved, more seem to appear in its place.

Judge Greene favors the deployment of a LEC database network in which the local exchange carriers could send all 0 plus calls to the interexchange carrier presubscribed at the billing address of the party accepting or otherwise being billed for the call. Thus, if a Boston user makes a call from a train station in New York, the Bell operating company would query its database to identify the billed party's home office carrier. It would then direct the call to that IXC's POP for completion – not to the carrier presubscribed from the telephone from which the call is placed. This is true to the concept of equal access and being able to choose a long distance carrier for the placement of all calls. A pay phone owner's presubscribed carrier would carry only those calls paid for with change at the time of the call.

Shopping for calling cards

With all of the different options offered by the carriers for calling card needs, it is hard for a telecommunications buyer to know that he or she is getting the best deal. Some things to consider:

1. **Watch for the deals.** The reality of multi-use cards is that the bank card portion of the card, though not unreasonable, is not competitive with the best bank cards on the market. AT&T's interest rate for the Universal Card is 8.9 percent above the prime rate (as of June 1, 1990, that totalled 18.9 percent) with no annual fee during the "charter period"; one of the best credit card rates in Connecticut was 13.5 percent with no annual fee. Also, AT&T's initial discounts on Universal Card calling card traffic, set at 10 percent in 1990, are not equal to what is available through other calling card options. SDN calling card usage is often discounted more than 25 percent.

2. **Watch out for alternative operator services.** Telecom managers should train employees always to check which long distance company is carrying the travel card call. AT&T in 1990 placed a recording on its network saying, "AT&T." Other carriers followed. This announcement is played right after hearing the initial tone. Listen for it. If a caller does not hear the tone,

either hang up and dial the equal access code plus the telephone number (e.g., 10288 + 0 + 203-645-0471) or wait for an operator to answer. If an operator comes on the line, ask for 1) a rates operator to verify the costs or 2) the carrier of choice.

3. **Watch out for pirates.** Calling card numbers are stolen as people use their cards. When dialing at a pay phone, users should cover the keypad with one hand to protect the security of the code or use card reader phones.

4. **Watch for cost cutting opportunities.** Some carriers have special deals which bypass surcharges on calls. MCI's Around Town option, for instance, allows for 950-access in a subscriber's home town without surcharges.

5. **Watch for 800 applications.** Many customers could replace calling card traffic with 800 traffic if a lot of the usage is back to the home office or another specific location. Users should analyze monthly bills for common calling patterns.

The Future of Calling Cards

Calling cards will take an increasingly important role in providing users with remote access to carrier networks, including all of the carrier's enhanced services and offerings. The trend towards integrating cards together to get "super cards" will continue as databases are linked and the necessary signaling systems are implemented.

Chapter 10

Facsimile Networks

One of the newest services available from the carriers is the facsimile broadcast network. These services are designed to take advantage of the explosive growth in the facsimile market. Simply, they allow a company to send fax messages to multiple points on a virtually simultaneous basis. This is ideal for sending out press releases or memos to all of your sales sites, etc. without tying up your fax machine for hours while the document is sent to one location after another.

The concept of facsimile networks was inaugurated by a company called USFaxsys in Philadelphia. The company reportedly ran into problems getting the network off the ground. It was MCI that became the first major carrier to successfully offer a fax network service, called MCI Fax. It was rolled out in late 1988. AT&T's fax product, AT&T Enhanced Fax, followed in late 1989. US Sprint's SprintFax and Cable and Wireless' Surefax have appeared recently, in 1990.

In addition to the interexchange fax network products, the local exchange carriers also are planning to get into the market. SNET's FaxWorks was the first offering in the arena, and the BOCs are planning such services as well. As of June, 1990, however, none of the BOC products had been launched.

Different Kinds of Fax Network Programs

There are two types of fax network services. Both have the same slate of basic features. The major difference between the networks is the manner in which users generate, update and maintain broadcast lists.

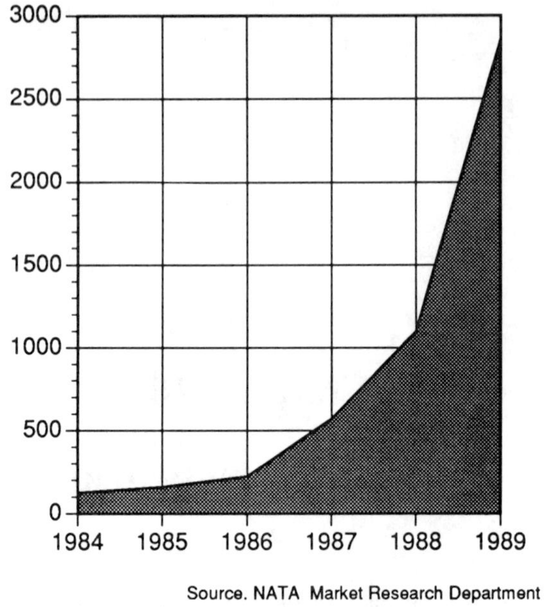

Figure 10.1 Number of facsimile machines installed in thousands.

Most fax network providers maintain the facsimile database for the customer. The carriers use the facsimile machine as the major link with the customer. Subscribers fill out specialized forms for the creation of new lists and to make changes to existing ones. These forms are then faxed to the carriers' main operations center for entry into the databases. This process generally takes about 24 hours.

This format of maintaining customer databases can be rather limiting because it inhibits spontaneous transmission to new numbers. Subscribers can bypass this 24-hour delay by punching in a stream of telephone numbers at the time of transmission to the fax network, but this can be a problem with long lists – mistakes can cause a customer to start the process over again.

The second form of database control used in facsimile networks is just starting to get off the ground – database access via personal computers. List maintenance is performed through carrier-provided software that is similar to most commercial mailing list maintenance programs. A subscriber makes any changes he or she wants to the database and transmits this to the fax network computer system, which updates the customer records accordingly. Then, the user can send the facsimile message using the latest fax network changes.

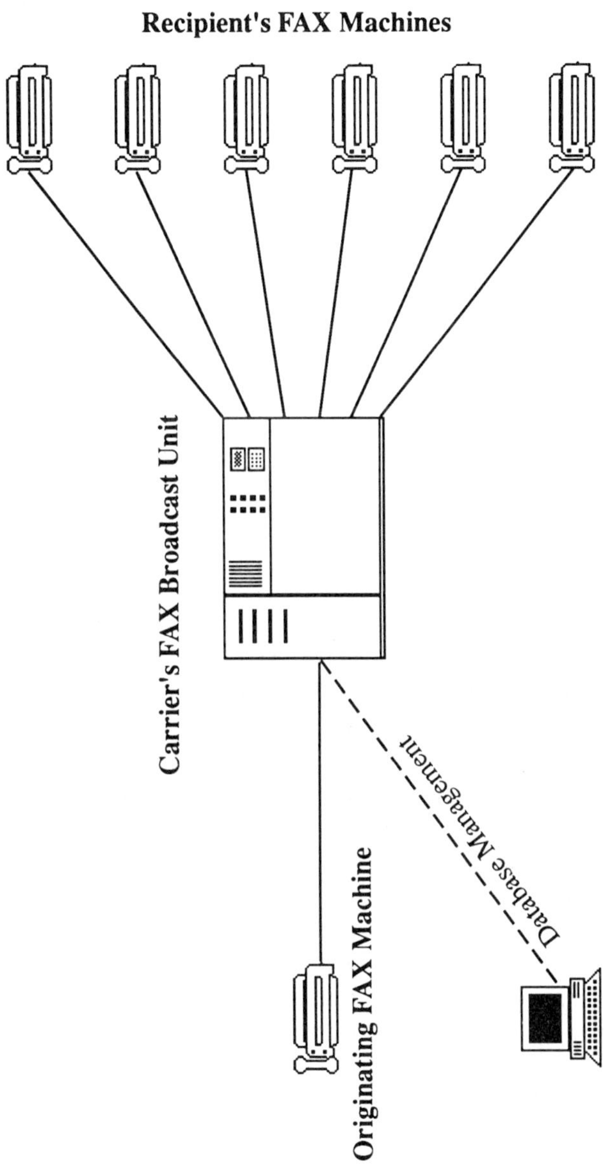

Figure 10.2 The path of a facsimile message over a facsimile network.

The PC based systems can be expensive, costing as much as $500 for the software, in addition to monthly subscription fees.

Sprint uses PC based software to manage their fax network. MCI offers both options. Cable and Wireless, AT&T and SNET perform database management for their customers using the first method.

How They Work

Facsimile networks are becoming increasingly user-friendly as enhancements are continually added based on customer feedback. Given the quick pace of the facsimile market, there has not been a great deal of time for the carriers to launch full-featured services. Like most other services, these fax network offerings are being introduced on a phased release basis.

Customers access a facsimile network in a number of ways. In the most general fashion, a customer simply send his or her facsimile message to a central carrier facsimile switch for temporary storage. At the time of the transmission, the customer enters or automatically transmits the coding for a predetermined broadcast list that is stored in the database of the carrier. The fax switch will accept the user ID number, designated fax distribution list or list of entered telephone numbers, and fax message and send the message to all of the addresses in the list.

At first glance, many customers might say, "My facsimile machine can do broadcasting too, so why would I want a fax network?" The key is transmission time. Most corporate facsimile machines have one dedicated line for sending and receiving messages. When a customer uses its own fax machine to broadcast thirty messages, most often these messages will be sent in a serial fashion, one after another. If the message takes ten minutes to transmit, the total transmission time is at least 300 minutes, or five hours. Not only does this tie up the machine for both inbound and outbound messages during that time, but there is a lengthy delay between the transmission of the first and the last message – five hours. This can be a problem for such time-sensitive transmission as press releases.

A fax network has multiple access lines for transmission and reception. Thus, it has the capability to send all thirty messages at the same time, or at least in an extremely narrow time frame. The user's fax machine is only tied up for the single ten-minute transmission to the network and the messages are received by the end points at approximately the same time.

The transmitting customer is charged a per-minute per-page charge for all of the calls which is slightly higher than standard voice-only per minute rates. The fax charges range from $0.35 to $0.75 per minute domestically. There might be service installation charges, software charges, or monthly charges as well.

The user maintains a high degree of control over the transmission of a message even though it is being performed by the fax network and not the users own fax machine. The carriers provide users with detailed reports on the delivery status of each message and when each transmission was completed. Most networks allow users to dial in and check on the status of a multisite transmission in mid-process.

How They Are Used

Obviously, the most popular capability of the fax network services is the ability to send multiple messages at one time. However, there are many other features that appeal to fax network users. These include:

1. **Timed international fax transmission.** MCI estimates that approximately 30 percent of all traffic over its fax networks is bound for international points. Many international sites are not in operation during the standard U.S. business day, due to time differences. The carriers' fax networks capitalize on this with store and forward features that will hold the transmission until the most cost-effective time period. In most cases, since the fax message arrives during the night, a delay of several hours to reduce costs will probably not be noticed.

2. **Fax mailbox service.** People who travel or who are in need of secure reception of fax messages can use fax mailbox services to store fax messages. For a monthly fee, a customer can get a fax mailbox into which all inbound fax messages are received. The customer can then access those messages from any fax machine by dialing an 800 number and entering the appropriate identification code. For added security, fax mailboxes are available with password codes that supplement the user identification codes. Only callers with the appropriate code combination will be given access to the fax messages in the mailbox.

3. **Fax information services.** For companies that want to use facsimile as a means to provide information for customer support or to prospective customers, fax networks have call-in distribution fax mail boxes that can be set up to return fax messages to customers that call an 800 or 900 number. Once reaching a sponsor's distribution box, the caller can access the infor-

mation he needs by pressing buttons on a touch-tone phone. Typically, a caller first enters the number of his or her facsimile machine and then presses buttons for all the information requested. Upon hanging up, the information is automatically sent by the fax network.

4. Transmission without a fax machine. Some of the fax network services will allow the subscriber to send fax messages from a personal computer without the need of a personal computer facsimile board. Sometimes this involves software provided by the carrier; other times, the carrier will allow customers to load ASCII files directly into its network through the use of off the shelf communications software, such as PRO COMM or XTALK.

Hunting for a Fax Network

Trying to select a facsimile network provider can be as confusing as buying regular long distance service. The first step in the fax network evaluation process should be a needs evaluation of one's fax requirements. Users should think about how they send facsimile messages and what that means in terms of requirements for the firm. Some specific things to think about when evaluating the different carriers are:

1. List maintenance. How important is it that a user keep control of the database at its premises? If a user is constantly juggling facsimile messages – adding new numbers all the time – then a PC based system might be required. On the other hand, if very few changes are needed, i.e., the messages are going to the same people most of the time, then the remote based option might suffice.

2. Costs. What are the per minute costs of using a fax network relative to using the site based facsimile machine? Fax networks are expensive – up to three times the cost of direct dialed transmissions. Subscribers should clearly define the times when the higher-priced facsimile networks should be used and when the direct, lower-priced home fax machine should be used.

3. Call rounding. How are the calls rounded? Some are full-minute rounded services, meaning that a 1.2 minute call would be rated at 2.0 minutes. Others are 6-second rounded services, so the 1.2 minute call would indeed be rated at 1.2 minutes. Potential users should keep rounding in mind when comparing the usage rates – the difference in cost between a 6-second service and a full-minute service is about 9 percent!

4. Database delay. What is the delay in updating the database once information is sent in via fax or directly via modem? Again, this gets to the question of how important it is to the subscriber to have the information updated quickly.

5. In-progress reporting. How accessible is the network to determine the status of the fax broadcast in the midst of transmission? Certain carriers do not allow the user access to that information – the subscriber must wait for the end of transmission report to find out if and when all faxes were delivered.

6. Usage reporting. What sort of reporting options are offered by the carriers? This can vary greatly. Some are available on a real-time basis, nearly on demand from the fax network. Others only come once a month with the bill. Companies should isolate the information needed and find out when it is available and from which carrier.

The Future of Facsimile Networks

The double-digit growth of facsimile usage assures the viability of facsimile networks as a product. One certain event in the near term will be decreasing usage prices for facsimile networks. The rates are artificially high right now, largely because of a lack of competition. All of the interexchange carriers and a number of local exchange companies will have competing products before the end of 1991, forcing the market price of the service down towards the price of an everyday voice call.

When this happens, usage of the networks themselves will blossom, possibly allowing more effective two-way communication via fax. It is not unrealistic to think that in the next few years, it will become popular to direct facsimile transmissions to a corporate network mailbox, not a facsimile machine. The network could then regulate the flow of messages to the facsimile machine. Alternatively, the facsimile machine could be programmed to dial into the network for messages on a regular basis.

The movement of the LECs toward voice mail offerings also suggests a prime market niche concerning fax – mailbox overflow delivery. When a fax machine line is busy the LEC network can intercept the call and re-route it to a fax mailbox. The fax network can attempt redelivery every few minutes, or users can call in to check for and receive overflow messages on demand. Note the similarities to LEC central office-based voice mail systems.

Facsimile machines require just as much management as other telecommunications network elements. With the increasing usage and costs of facsimile, fax networks provide a sensible way to start better management of the system.

Chapter 11

900 Services

One of the more interesting segments of the long distance industry to watch in the 1990s will be 900 services. The general public is just starting to accept these "paid-for" services. The industry is growing out of the association with limited applications like 1-900-VAMPIRE and 1-900-LOVELINE, and moving towards more respectable corporate applications, such as retail marketing and information distribution.

900 service draws its name from the number prefix used in accessing the service – just like 800 or 700 services. The most significant single fact about 900 services is that they are "pay per call" services – the caller is charged each time he or she dials the 900 number. This is the opposite of 800 services in which the called party pays for the call. 900 service charges are included with the monthly local telephone bill.

AT&T launched 900 services more than ten years ago as a means to generate profits for the information industry. Originally called "AT&T Dial-It 900 Services," their outlook was considered quite promising. AT&T hoped that the nation's businesses would take to 900 services as they had to 800 services – and that would mean sizable revenues for the long distance giant.

900 services started with a bang. The first application was both noteworthy and visible – it allowed callers to vote for the winner of the initial debates between President Jimmy Carter and Governor Ronald Reagan. More than 500,000 callers phoned over three evenings to give the first hints of who would win the next presidential election.

Since that time, 900 services have grown considerably in popularity. Today it has $300 million in retail sales; by 1992, the figure should near $2 billion. Initial product offerings were limited since the available services only allowed "passive" programming. This was one-way communication – or listening in on messages. In addition, there were no opportunities to interconnect a customer's own equipment with the carrier's network.

For these reasons, serious business applications were conspicuously absent from 900 services in the early years. Most of the revenues were generated by less respectable programming, which today's 900 services reflect, offering opportunities to get the latest sports scores, horoscopes, or pornographic taped messages.

As a corporate tool, 900 is now starting to get off the ground. Corporations have been slow to catch on because corporate managers have not yet learned how to make 900 services fit with their corporate telecommunications strategies. It was not until the spring of 1989 that interactive 900 services hit the market. Interactive services are differentiated from their passive counterparts by the use of touch tone by the user to interact with the network and programming. With the new services came new equipment and connection options that allowed for the first time corporate applications over 900 services.

The slow development of corporate 900 can be traced to slow awareness of new 900 developments. Another factor is that some corporate telecom managers have been leery of implementing large scale 900 programs due to public association of 900 with pornography and other questionable programming – much of which is associated with 900's local exchange company counterpart, 976 services. In the most successful corporate uses of 900, this image was overcome by the reputation and innocence of the services and providers themselves. The National Aeronautics and Space Administration (NASA), for instance, had a hugely successful program in which callers could listen to communications between NASA and its space shuttles.

Slowly and cautiously, corporations have been adding 900 to their list of revenue sources. Each month, applications are found to make use of these profitable services, but not without a lot of failures in the process. Setting up a 900 application is tricky and wrought with potential catastrophe if managers are not careful.

The Principle behind 900 Services

900 services are also commonly referred to as "audiotex" services. Companies that offer 900-number programs are generally referred to as "information providers" or "sponsors."

The 900 services industry is based on the simultaneous access of thousands of callers to the same message. Information providers supply any of four major types of audiotex services: passive, interactive, polling and conferencing. 900 services may be

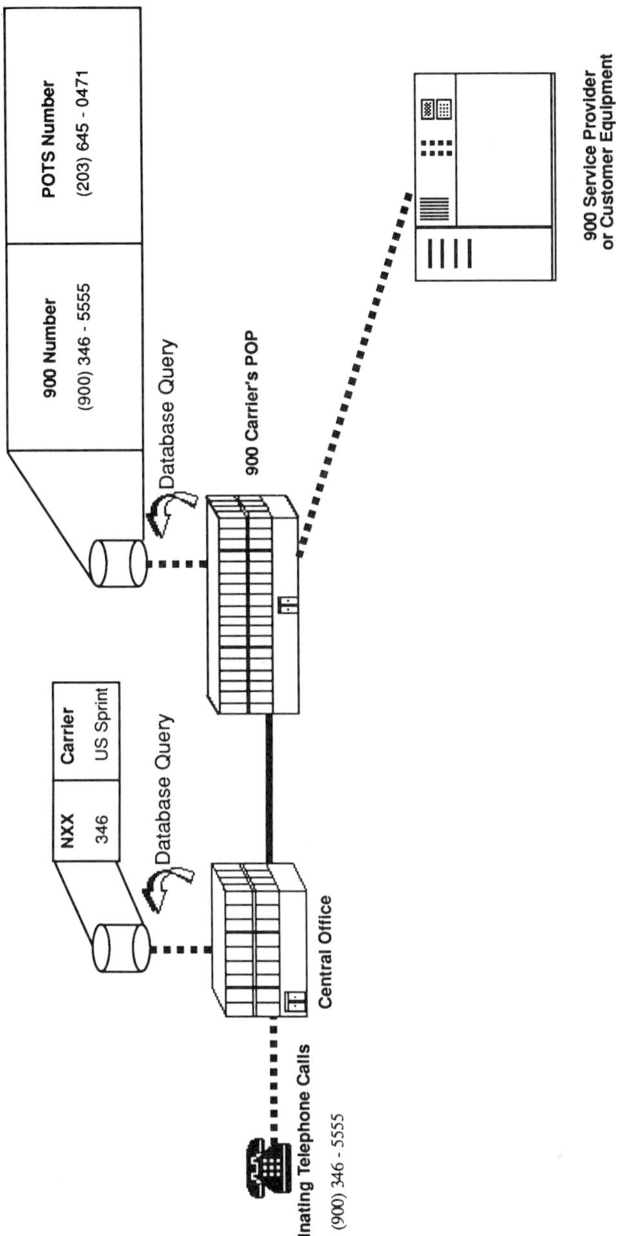

Figure 11.1 The routing of a 900 Call.

passive or interactive, depending on whether the program is designed to take input from the caller. Passive applications are typically limited to information distribution programs, while interactive applications often require participation with the caller.

The 900 services may also simply poll callers for one or two responses, typically "yes" or "no." The services may also link callers for simultaneous conversation – the so-called GAB (Group Access Bridging) lines. These are the love lines, romance lines, and other "barge in" lines.

The 900 Network

Unlike the majority of 800 numbers, 900 numbers are constantly changing hands. Many times, programs are one-time applications and, when done, the number shifts to another customer or application. For this reason, no formal public 900 directory exists. Some 900 numbers, such as the Wall Street Journal's 900-JOURNAL or USA Today's 900 weather line, are more permanent.

To access a 900 program, callers simply dial 1-900 plus a seven-digit number. As with 800 calls, when the 900 number passes through the local central office (CO), the switch screens the first six digits (900-NXX) and routes the call to long distance carrier to which the NXX is assigned. The interexchange carrier transports the call to the information provider's processing equipment – whether owned by the information provider, the service bureau, or the carrier.

900 Image Problems

900 is a developing technology. Many people do not know what 900 numbers are. A recent survey of 12,000 people by Advertising Research Corporation and Strategic Telemedia of New York City showed that 55 percent of this nation's callers had never heard of a 900 number.

Awareness of the medium is lacking. Of those who do know what a 900 number is, most are confused as to what exactly the service provides. Ninety-five percent of the nation has never dialed a 900 number. Those who use 900 numbers on a regular basis are unable to associate a unique purpose with the service, although polling and personal entertainment are mentioned frequently.

In some ways, the 900 industry is recovering from an early bad image. Much of the initial programming was less than reputable. Low entry barriers, lack of program supervision or policing and a fast money image brought a flurry of "get rich quick" entrepreneurs into the industry. The first programming concentrated on pornogra-

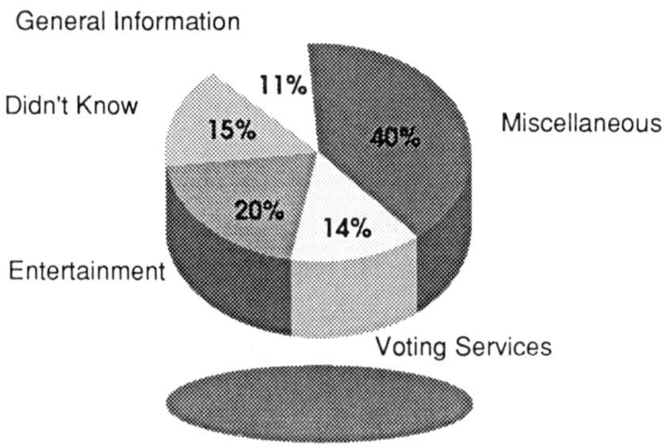

Specific Breakdowns of Different Catagories

Personal Entertainment:*	11% Dating Services
	4 % Sexual
	4 % Rock Star Updates
	2 % Story Lines
	2 % Joke Lines
	1 % Wrestling Updates
General Information Services:	3 % Sports
	2 % Horoscopes
	1 % Soap Opera Updates
	5 % Other
Miscellaneous:	40 % Interaction with TV Shows

* Multiple Selection Question

Source: Strategic Telemedia

Figure 11.2 How people perceive 900 numbers are used.

phy, group talk, and children's topics – each of which brought public scorn in one way or another, but were immensely profitable.

Corporations installing 900 numbers today will have to cope with this reputation, as well as other perceptual barriers. Companies looking to switch 800 lines to paid-for 900 lines, for instance, can upset customers and lost sales could easily surpass potential 900 profits. It also sends a signal to customers that the company is now charging for something that has been free in the past. This can convey a "cheap" image. The same is true of those companies that wish merely to pass the burden of sales and information calls to the customer; while the company may reflect reduced

900 Consumer Attitudes

Toll-free numbers are continuing to burrow themselves into our lifestyles. A 1990 study, conducted by Advertising Research Corporation and Strategic Telemedia, showed that more than 39 percent of the nation's population use an 800 number every week, and more than 50 percent call toll-free every other week. More than 88 percent of the nation's population use 800 numbers on a regular basis. "Frequent 800 callers" – those that placed an 800 call within the past two weeks – averaged four 800 calls per week. The study surveyed 12,000 customers in ten markets during 1989-1990.

Interestingly, there was a pronounced skew towards upper income households in terms of the number of weekly calls to 800 numbers. Overall, calls for the expressed purpose of purchasing an item make up a fifth of all 800 calls. But in the frequent caller category, purchases over the telephone via 800 are more than double the average. The study estimated that these frequent callers represent more than half of the $5.5 billion 800 services market.

The widespread awareness and usage of 800 has yet to hit 900 services. 55 percent of those surveyed had never heard of a 900 number. Only 5 percent had ever called a 900 number. That means 95 percent of the nation has never called a 900 number.

In fact, there appeared to be a lot of confusion as to what exactly a 900 number was. An interesting indicator of this present perception of 900 usage the aware non-user category – those people who are aware that 900 numbers exist, but do not call them.

One thing appeared to contradict conventional wisdom – only a small percentage saw the perceived purpose of 900 numbers as the carriage of sexual messages.

Will 900 take the place of 800 in many applications? Industry experts say that few consumers will migrate to 900 services unless perceived benefits exist. The ARC and Strategic Telemedia study found that, of the frequent 800 number caller group, 33 percent expressed willingness to pay as much as $2 for their most recent 800 call. These users had found certain applications over 800 lines that they had grown to appreciate and therefore were willing to pay for. However, among the aware non-user group, one out of three people do not call 900 numbers based on the fact that the call will cost money.

product costs from the switch in lower market prices, the consumer typically will not make the association with 900.

Why 900?

So why would a company want 900? 900 services have a lot going for them. Despite the problems, there is a lot of money to be made in the 900 business. They are easy to use. All but a few percent of the nation's population owns or has ready access to a phone, and most are familiar with how they work too. People are becoming quite used to interactive phone technologies, particularly since the explosion of voice mail and automated attendant equipment into corporate America. While shop at home and videotex networks for personal computers continue to experience slow growth, 900 services are taking off.

The applications for 900 are diverse. Many companies may be drawn to 900 services as a replacement for 800 service since 900 shifts the cost burden to the caller. Customers may have access to an 800 number during their warranty period, and a 900 number thereafter.

Other companies look to 900 to provide marketing information about consumers. 900 services are playing an increasingly important role in television programming, for instance. Some applications have been tongue in cheek – voting whether Larry the Lobster should be boiled or saved on Saturday Night Live. Others control actual content. In late 1987, Music Television (MTV) first used 900 service as a replacement for postcards because there was too little time for the mail to reach the station. MTV was overwhelmed when nearly one million callers dialed the 900 number. Since then, 900 numbers have been used frequently to decide which videos should be shown and when. Companies also can use 900 to offer enhancements to products or services, with minimal capital outlays. A turkey producer could offer customers recipes over 900 numbers; Nintendo consumers could dial for the latest strategies for new games.

Moreover, 900 could be mixed with other existing technologies to yield significant added value to callers. An amateur auto mechanic could call a 900 number for help in repairing a car. He or she could listen to messages, working through various menus until the appropriate information is found. Then, full supporting diagrams and instructions for repair could be returned to the caller via facsimile. The same idea could be copied to almost any consumer product.

In these ways, 900 serves to bring the consumer closer to the product. Direct interaction of this sort can bolster customer loyalty and increase corporate presence. According to television gurus, MTV's use of 900 to control programming has enhanced viewer involvement with the show and has created more loyalty. This, in turn, generates profits when selling advertising space because MTV can point to such audience involvement.

Psychologists agree. Many experts specializing in human interaction and reception of new technologies have shown that people feel better when they control their own destinies and can interact, even when these "relationships" are over the telephone. This explains why recorded outbound telemarketing systems have been so dismally received, but interactive 900 services are welcomed.

Finally, 900 services provide important marketing indicators for service and product development. By tracking customer response through 900, companies are able to combine revenue generation with product development. As with 800, captured automatic number identification allows each call to be matched with a database of names and addresses to build targeted mailing lists.

Increasing Corporate Use

900 services are gradually gaining acceptance as a valid medium for business relations. Credit also must be given to the carriers for pushing the market in this direction. Sprint Gateways (a subsidiary of US Sprint) and MCI, for instance, will not accept any "adult," "GAB" or other programming that would reflect unfavorably upon the company. Other carriers are following suit.

One of the most visible companies using 900 today on a regular basis is USA Today, published by Gannett. USA Today regularly uses 900 numbers to stay in touch with its audience and to poll readers about information in each issue. On a typical day USA Today may advertise several 900 numbers. A recent issue had four different 900 numbers, that offered sports scores by phone; asked readers to vote on the athlete of the year (Football star Joe Montana was winning after 2100+ calls); offered horse racing results from 90 Thoroughbred and harness tracks in North America; and provided weather from around the nation. Other companies have been similarly successful in prompting readers, callers, owners, purchasers, etc. to interact with them via 900.

900 Case Study: Prime Time to End Hunger

Perhaps one of the most ambitious 900 programs to date was the December, 1989 End Hunger Network (EHN) campaign. During the month of December, the nation's three largest television networks – ABC, CBS and NBC – combined their collective strength to help build public awareness about hunger and the homeless.

From December 1 through 20, *The Cosby Show* and *Cheers* (NBC), *Murder She Wrote* and *Designing Women* (CBS), and *thirtysomething* and *Head of the Class* (ABC) aired single episodes written around the hunger and homeless issues. At the end of each show, the actors appeared in two–minute segments urging viewers to call to find out how they could volunteer in their areas. The goal was to increase public awareness and subscribe volunteers – not to raise money.

The supporting 900 network was enormous. These six shows reach 290 million viewers each week. The 900 network that resulted was capable of handling up to 10 million calls during the three-week campaign. No single carrier was expected to be able to handle the traffic volume during peak hours, so the EHN used AT&T, MCI, Sprint Gateways, and Telesphere to provide the 900 call transport. Call processing was supported by more than 20 service bureaus around the country.

The viewer dialed a 900 number shown on the screen. After listening to a message about volunteer opportunities, callers were prompted for their name and address. Each caller's spoken information was transferred from computer audio storage onto audio cassettes and shipped to a group of transcribing organizations. From there, the information was sent to a printing company where personalized self-mailing letters were generated. These letters were sent to callers with the phone number and other information about volunteer centers in the caller's area. Time from call to mail delivery was approximately nine days. The 900 call cost callers approximately $2 and revenues from the calls supported the entire campaign.

In the end, much of the contingency planning was never used. Despite the grandiose plans of the EHN organizers, only 100,000-200,000 calls were received. Nevertheless, the visibility and coverage for 900 services as a fund-raising medium helped the industry in general.

Carrier Services

Presently, there are only five providers of 900 service – AT&T, MCI, Sprint Gateways, Telesphere and Allnet. Allnet's product is a regional service covering five Mid-Western states.

The carriers offer two ranges of 900 services – basic transport services and enhanced processing services. The five carriers each offer a basic 900 service that provides for the transport of the call through the interexchange network, the billing of that call either by carrier itself or through the local exchange carriers, and the collection of the revenues from the local telephone companies. Then, each of the carriers offers some sort of enhanced 900 service that also includes the equipment and processing of the 900 program. Enhanced processing offerings include the programming of the message, the provision of the playback equipment, and any special handling of the calls as they enter the carrier's domain.

US Sprint established its Sprint Gateways division to market these basic transport and enhanced processing services in the 900 industry. AT&T has started a joint venture with American Express Information Services Co., called Call Interactive, Inc., to provide interactive voice services. The carrier also is gearing up its Enhanced Services Complex, which will provide for simple storage of messaging and transport of the call. Telesphere and Allnet both use Ameritech 900 Services for provision of enhanced processing. MCI uses several service bureaus for the distribution of its basic 900 service.

AT&T has the oldest product, built around its Dial-It 900 service. They renamed the entire product line MultiQuest in 1989. At present, there are four AT&T services:

1. **MultiQuest.** An interactive service which allows customers to link a company with the caller through 900 numbers (up to 1800 calls per min.).

2. **MultiQuest Hi-Cap.** Same as MultiQuest, but designed for high-volume (up to 7500 calls per min.).

3. **MultiQuest Broadcaster.** A passive service which allows recorded and live information in a broadcast format.

4. **MultiQuest Call Counter.** Allows sponsors to poll callers.

900 Case Study: Fox Broadcasting

The proliferation of 900 services into the television domain is increasing. The Fox Television Network has been integrating 900 numbers with its shows to increase viewer participation.

Consider the show, *Married With Children. Married With Children* depicts a rather unusual couple in a standard love-hate sitcom relationship. Toward the end of the valentine show last year, Peg, the wife, tells Al, the husband, that she loves him. His valentine's gift to her would be a returned "I love you, Peg." Fox ended the show a few minutes early and polled viewers as to whether Al should say "I love you, Peg" or "Not tonight, Peg." Producers filmed both endings.

More than 750,000 calls came in as the network aired the remaining 2.5 hours of Sunday shows broadcast by Fox Television. Afterwards, Fox played the sappy vote winner – "I love you, Peg."

The show's February 14 rating – a 7.0 share, up from 4.7. The next two Sunday's saw similar boosts for all Sunday programming. Moreover, by holding the ending until after 2.5 hours of other programming, viewers were exposed to shows they did not usually watch.

Sprint Gateways offers three services:

1. **Voice 900.** Provides interactive and passive information distribution through Sprint Gateway's equipment.

2. **Network 900.** Provides the transport, billing and collection of a 900 call; processing is left up to the customer.

3. **Data 900.** For computer-to-computer applications.

MCI, as with its other long distance services, has chosen to market all of its 900 options under a single name, MCI 900. This contains all of the passive and interactive information services, as well as polling and broadcast features.

Company	Product	Application Type
Allnet	Market Research (Network Billing Option)	Passive and interactive information, polling
AT&T	MultiQuest	Passive and interactive information, polling
	MultiQuest Hi-Cap	Passive and interactive information, polling
	MultiQuest Broad-cast	Passive information
	MultiQuest Call Counter	Polling
MCI	MCI 900	Passive and interactive information, polling
Sprint Gateways	Network 900	Passive and interactive information, polling
	Delta 900	Passive and interactive information, polling
Telesphere	National 900	Passive and interactive information, polling

Figure 11.3 Available 900 Services.

Telesphere has three services – National 900, Dynamic 900 and Data-Cap 900. National 900 is the base service. Dynamic 900 and Data-Cap 900 provide enhanced call processing.

Allnet has two service options under its main service, MarketReach 900. One is the basic transport, billing and collection service, and the other adds processing functions through Ameritech 900 Services.

Service Bureaus

The long distance carriers are not the sole source of 900. They are outnumbered by more than 100 non-carrier service bureaus that specialize in providing enhanced 900 service, based on the resold carrier 900 basic transport services. In fact, the vast majority of current information providers use a service bureau for processing 900 programs. Few customers are themselves set up to interface directly with the long distance carrier basic transport services. While the carriers are trying hard to move into the service bureau market with their enhanced processing services, they now sell predominantly transport, billing and collection services to the service bureaus.

In many ways, service bureaus are far better positioned to sell 900 than the carriers – they have all the experience. Many service bureaus have been selling 800 services for years – selling 900 is a natural evolution. After all, 900 is just "paid for" 800 service.

The service bureaus resell the five carrier 900 services. Typically, the bureaus are reselling the basic transport, billing and collection services of the carriers – not the call processing enhanced services, which they provide themselves. The service bureaus already have the equipment in place to provide these features.

Most service bureaus resell more than one carrier's services. In fact, it is not uncommon to see firms selling each of the five carrier services. For any particular application, the bureau will pick and choose among the existing service providers and in-house equipment to get the best combination of end to end functionality. This is much the same process in designing an 800 application.

Perhaps the biggest selling point of the service bureaus is their experience and comfort in dealing with both 800 and 900. For years, these companies have specialized in incoming call management, including all of the application-specific and custom-design networking that this implies. The level of service is unrivaled by any of the carriers, and this is why most sponsors go to a service bureau for their application. At the same time, this makes it extremely difficult to compare service bureaus with one another. It is like comparing advertising firms – while their past campaign successes are important, it is the individual application at hand that matters.

Choosing a 900 Carrier

Getting started in 900 involves picking a carrier and processing entity, if the company does not have equipment of its own. These are not necessarily independent decisions – picking a carrier will most likely limit the choice of service bureaus to a subset within the industry, and picking a service bureau can often limit the choice of carriers.

Choosing a long distance carrier for a 900 service is often a process of elimination – deciding which carriers cannot do what is needed. Among the more important factors to know about when picking a carrier are:

1. **Program content.** Perhaps the most important factor, the carriers are clamping down on children's and adult programming. If the application involves one of these two types of 900 service, certain carriers will not carry the calls. Sprint Gateways and MCI both closely scrutinize all such applications, for instance, and may decline to offer service. Along the same lines, AT&T has set up premium billing guidelines, where each program must meet these criteria to be accepted. AT&T is also working with CARU (Childrens Advertising Review Unit, of the Better Business Bureaus) to provide guidelines for childrens programming; specifically, a $4 limit per call and call rate in advertising must be presented on screen as well as verbally. Likewise, MCI has extablished guidelines for billing and collection of 900 services.

2. **Processing Time.** 900 services can take only days to set up through a service bureau. In many instances, as little as five days can be required. However, the time it takes for a carrier to initiate service can vary. Typically, Sprint Gateways, MCI, Telesphere and Allnet take less than two weeks; AT&T takes two weeks or more. If time is a critical issue, AT&T may not be a viable choice.

3. **Capacity.** If there is one figure most misquoted in the 900 industry, it is capacity. In most sales literature, the numbers quoted are meaningless because the stipulations for capacity calculations are so outlandish. In the extreme, a carrier could say "If there is no other traffic on my network, then I could complete X number of calls." How much capacity is needed is directly related to the application, whether it is polling, interactive communi-

cation, GAB lines, etc. In most cases, large capacity is important for high-volume response applications such as polling. Capacity can be a local constraint as well as a network constraint. If local access is not available then network bottlenecks may not even come into play because the traffic may be blocked at the local level. Realize, capacity is rarely an issue, very few programs will threaten the integrity of a carrier's network. However, if a sponsor expects exceedingly high traffic volume in a short period of time, then certain carriers may not be options. High-volume applications require specialized solutions. A carrier may have to link together several service bureaus to achieve the desired throughput (see 900 Case Study: End Hunger Network). Companies should be sure to get firm commitments from the carriers because this always tends to be a grey area.

4. Equal Access. Surprisingly, equal access is a major issue with 900 services. Without equal access, a carrier does not receive automatic number identification with the calls. In absence of this information, the carrier must have a recording agreement with the local telephone company, or the carrier cannot bill for the calls.

AT&T, because it has direct connections to all end offices, holds a decided advantage over the other carriers who must wait for feature group D (equal access) to become available at each end office to provide 900 services out of that CO. This may or may not be a consideration for a particular application depending on an information provider's target audience.

According to AT&T, more than 23 percent of the nation's exchanges are not yet converted to equal access. The standard answer from AT&T's competitors is that these exchanges only cover the least populated portions of the US; the vast majority of callers are covered by equal access." This is true to an extent. Looking at pure converted exchange totals does not address the distribution of households. However, some carriers do not yet have ubiquitous access, and therefore this should be considered.

AT&T's competitors can accept calls from the non-equal access locations, but, depending on what agreements are in place with the local telephone company, they may not be able to bill for the calls. Each carrier deals with the problem differently – Sprint Gateways will accept non-billable calls and

Selecting Vanity Numbers for 900 Service

As with 800 service, 900 numbers can tell the caller a great deal about a company's product before the call is even placed. For instance, the Wall Street Journal has 1-900-JOURNAL for calls to its stock information line. Here is a list of how Bellcore has assigned NXXs for 900 service:

Allnet:
224 753

AT&T:
342 344 370 420 740 773
786 820 860

MCI :
226 289 446 476 622 678
776 988 990

Sprint Gateways:
230 246 346 386 463 468
535 568 847

Telesphere:
234 258 321 369 456 654 741 872 963 999

Figure 11.4 By selecting the right arrangement of numbers, companies can get an easy-to-remember number so their callers do not have to look up the number all the time.

absorbs the costs as part of its operating expenses; Telesphere, Allnet and MCI may block the calls.

Realize that these billing issues are transparent to the user. Callers to 900 lines pay a published per minute rate to the local telephone companies that covers the long distance and program charges. The telephone companies then compensate the long distance carrier a specified amount of the proceedings, and the carrier sends the sponsor his or her appropriate portion.

5. Vanity phone numbers. Since the local exchange carriers can only provide six-digit screening, each carrier can sell only those exchanges allocated to it by Bell Communications Research (Bellcore). If a sponsor is looking for a particular telephone number, then that could define the choice of carrier.

6. Pricing flexibility. Carrier pricing schemes vary. AT&T is the only carrier that has fixed pricing per 900 exchange; that is, AT&T has a pre-determined price for each of its allocated NXXs. All of the other carriers provide flexible pricing, regardless of the exchange. With AT&T, a customer has to change the telephone number if he or she wants to change the rate charged. AT&T says that it intends to change this pricing limitation sometime in 1990.

7. Disbursement of funds. How long does it take the carrier to pay out funds from 900 services? They will say 60 days, but often reimbursement can take up to 90 days with some carriers. While this may not help a sponsor choose among carriers, it is important in determining cash flow needs. Companies should check out their written terms and agreements, and get the carrier to put their disbursement period into writing.

8. Carrier billing guidelines. Some carriers have rate ceilings that may be charged on any particular call. If a sponsor wants to bill $50 per call charges, and the carrier has a $10 per call rate cap, then that carrier may not be an appropriate supplier.

9. Intrastate calling limitations. Certain carriers, notably AT&T, may have problems completing calls when the caller is in the same state as the information processor. Calls may end up being blocked for much of the target audience. Companies should check the intrastate carriage capabilities of each carrier in the state.

10. Specialized management reporting. Some carriers provide more enhanced call detail and traffic analysis reports, giving information providers more detailed information about who is calling and from where. Sponsors should ask for sample reports and custom billing options.

In the end, pricing between the different carriers often matters very little in the overall decision. If after all other factors are taken into consideration there are still multiple carrier services left, then pricing becomes a factor. However, rarely does price alone determine which carrier a customer will use for the transport portion of the 900 call.

Choosing a Service Bureau

Service bureaus themselves merely add processing features and options to the core carrier services. A typical service bureau can manipulate 900 products in thousands of ways to suit particular applications. There are a number of key categories that a future information provider can check to start deciding which service bureau to use. Among them are:

1. **Competitive potential.** Is the service bureau a possible competitor? If a service bureau has its own 900 programs running, are they similar enough to the sponsor's own programs to cause problems?

2. **Ownership of 900 lines.** Does the service bureau or information provider own the 900 lines? Some service bureaus rent companies their own numbers. This can be a dangerous situation if a program is successful and the service bureau starts raising prices or decides to provide the 900 program itself.

3. **Carrier services available.** Which carriers are available from the service bureau? If only one, is this the carrier that can best suit your needs? The broader the selection of carriers, the more flexibility the sponsor will have in the long run. Realize, however, that the ability to switch among different carrier services is limited – until the LECs develop the appropriate databases, customers have to change 900 numbers to switch carriers. (Note: a sponsor does not have to change its 900 number to switch service bureaus if the sponsor owns rights to the 900 number.)

4. **Access options.** Does the service bureau require full T-1 access or can it support individual dedicated access links as well? Can the firm provide business line access?

5. **Equipment capability.** 900 services are changing products. Information providers are constantly adjusting services to better meet the needs of the marketplace, and service bureaus need to be outfitted well enough to handle

this ongoing change. Would-be sponsors should compare each bureau's equipment inventory to make sure future needs can be met.

6. Use of owned equipment. Service bureaus make money by providing sponsors with in-house equipment. However, some sponsors want to use their own equipment. Can the information provider supply its own equipment? Some service bureaus limit the amount of equipment that may be brought on site, and many charge extra for this privilege.

7. Equipment manufacturer. Is the service bureau tied to a particular equipment vendor, either directly or indirectly? A sponsor company does not want to be tied to any one vendor's equipment – this limits the ability to switch service bureaus.

8. Transfer capability. Can the service bureau transfer inbound calls to another site? Many sponsors use service bureaus for the preliminary processing of a call before forwarding it to the sponsor's site for further processing. Some service bureau firms only have inbound capabilities. Without adequate outbound facilities as well, an information provider may not be able to route calls to live operators at his or her own site.

9. Storage limitations. What are the limitations of a particular service bureau's equipment when it comes to storing messages? Are there extra charges for storing more than a certain number of recordings?

10. Location of processing centers. The location of the service bureau's processing center can affect the taxes that apply to each call, and therefore is critical in comparing service bureau costs. Any call that begins and ends in the same state is typically subject to a whole range of other state and local taxes, which do not affect calls terminating outside the state. Sponsors should look for service bureaus that have multiple processing centers, with at least one outside the state from which the most calls are expected.

11. Optional services. What extra services does the service bureau have that would enhance the 900 service application? Does the service bureau provide reverse-identification services, that is the matching of the captured automatic number identification of each call with a database of names and addresses to build mailing lists? How about name and address collection from recordings? If yes, how fast is the turnaround from recording to transcription? Another service to look for is call tracking – the ability to follow a call through the system for market research purposes? Another feature to check for is documents on demand – receiving hardcopy information immediately via fax.

12. Cost factors. There are several standard cost factors to examine. These include monthly service fees, additional per minute charges, volume discounts, additional program charges, and any other extra charges that may exist.

13. Length of Campaign. Some companies may be struggling with the service bureau or in-house processing question. In most cases, companies that are thinking about implementing a 900 program for a short-term promotion would not want to buy their own equipment; the service bureaus have invested many years and dollars in their 900 facilities and it would not be worth the effort to duplicate those services for a short-run 900 offering. However, on a long-term basis, companies might indeed want to consider buying their own equipment sometime in the future. Most industry experts would recommend trying the service with a service bureau first, keeping in mind the above note on owning the rights to the 900 number.

LEC Billing Problems

Having selected a service bureau and carrier, there are still a lot of other issues to confront. Billing has and will continue to be a big problem with 900 services.

Local exchange carriers are not willing to bill for some of the larger revenue producing programs for fear of collection and image problems. Many limit the amount of money a carrier may collect for a given call. Others will not bill for calls at all. Largely, the LECs are not interested in 900 numbers except for specific applications. These tend to be the most straightforward programs – weather, sports, horoscopes. Among the programs for which the LECs will not generally bill are fund raising and merchandising, which is where the large future revenues lie. Many LEC decision makers fear 900 merchandising services which charge customers for purchases through the monthly telephone bill. Irate customers who do not like the goods or service or the sponsor might refuse to pay the bill from the phone company. Such problems create potentially large collection and customer relations problems for LECs.

900 Services To Bypass Automatic Number Identification

There is at least one way to get around the caller number identification mentioned in Chapter 10. Private Lines Inc. of Beverly Hills, CA has come up with two separate 900 numbers that users can call when they wish to block the transmission of their originating numbers to the called party. Callers dial the 900 numbers and get a network tone. They then enter the number they wish to call. Subscribers to caller ID services – including those of the interexchange carriers – see nothing more than zeros on their display screens.

The two numbers cover domestic and international calling respectively. Calls to the domestic 1-900-STOPPER are billed $2 per minute and calls to the international 1-900-RUNWELL cost $5 per minute. Charges show up on the monthly local telephone bill.

The long distance carriers also suffer from the LECs' difficulties with collection. The carriers must set aside a portion of their 900 revenues to cover bad debt because of expected collection problems. Adding to the initial difficulties with remittance are the limited capabilities of the LEC billing systems. Some of the carriers cannot specify what bad debt applies to which program because they do not get that information from the LECs–they just get a bulk monthly bill. The carriers seek to limit their bad debt exposure by controlling program content. Historically, GAB lines have some of the highest bad debt.

AT&T has an edge on its competitors because the carrier provides its own account inquiry services. Thus, AT&T deals directly with customers concerning questions about billing. With this in mind, AT&T can choose which services it wants to support – even merchandising. Other carriers – reliant on the LECs to handle the billing inquiries – cannot. The fact that AT&T performs its own billing inquiry service is significant. Most LECs have the standard policy of refunding any questioned 900 charges. AT&T, on the other hand, evaluates billing questions on a case by case basis. This accounts for the larger amount of bad debt incurred by AT&T's competitors. AT&T, through its evaluation process, has the ability to control its own losses.

The 900 industry must fight an uphill battle on the collection front as telephone subscribers continue to complain about responsibility for 900 calls on their monthly bills. As a result, more than fifteen states have passed 900 related legislation and other states have such discussion on the agenda.

900 Blocking

Another major factor hindering the corporate 900 explosion is blocking, by both users and local exchange companies. Regulators and local exchange companies around the country feel the easiest way to get rid of bad debt in the industry is not to allow the calls in the first place, so they block the calls.

The broadest means of blocking calls is at the central office level. Many independent telephone companies automatically block all 900 calls from their territories. These carriers do not want to deal with potential bad debt and image problems. Local telephone companies are also providing the means for individual customers – residential or business – to block 900 calls from their phone lines. As of June, 1990, 16 states have approved 900 call blocking and many more are considering the proposition. One of the more recent actions, from Connecticut's Southern New England Telephone (SNET), makes blocking available only to residential customers. The blocking will not apply to non-900 local exchange company "infotainment" lines, which typically start with the prefix 976.

With LEC 900 blocking, all 900 calls from the affected line are blocked by the local telephone company as the call is screened in the local central office. Thus, in the absence of six-digit or ten-digit screening of these 900 calls, it is impossible to separate the true business oriented 900 calls from the "Dial-A-Joke" and "Personal Romance Hotline" numbers.

Also, many companies are blocking all 900 prefix numbers as they pass through the PBX. If a company employee needs to make a business related 900 call, typically the call is billed to a corporate credit card and validated on expense reports.

In the future, 900 blocking may become more selective in nature. Customers have the means to provide the most control over 900 through the PBX. Enterprising telecom managers could decide not to use LEC 900 blocking and instead control the blocking in PBXs through three- and ten-digit screening. Thus, selective 900 numbers – i.e., corporately approved numbers – could be passed through to the local telephone company for completion. Non-approved numbers would be blocked. This requires a considerable amount of work and would implicitly rule out spontaneous 900 traffic. It would be nice if this same function could occur through the central office, but the local exchange carriers are not planning to offer the capability anytime soon.

The Future of 900

There is a lot of room for the enhancement of 900 services. One example is through automatic fax return – the ability of a 900 number to return a hard copy of information to a fax machine at the end of a message. With the proliferation of fax machines nationwide, this service is just starting to take off.

Another emerging 900 service is market research information. Telesphere, for instance, recently launched its Prizm 900 service that combines many diverse databases to create a complete analysis of callers to its 900 programs. Thus, information providers can receive a breakdown of all callers by geographic location, psychological profile, and demographic standing, to get a broad characterization of the average 900 caller. This information is invaluable in designing targeted advertising.

Competition in the industry is keen. To be successful, service bureaus and carriers will become more innovative in helping the information providers themselves differentiate their 900 numbers in the marketplace. Sprint is building into its 900 service merchandising options that would allow products and services to be ordered via 900 and processed by Sprint. The billing for the 900 call would be on the local telephone bill, but the merchandise would be billed on the credit card invoice. Sprint hopes this system will suppress LEC fears of 900 and merchandising.

For end users, the challenge will be how to best market 900 programs. Most companies are far away from building a complete marketing and sales program around individual 900 programs. Telecom managers will have to cope with integrating 900 with existing 800 and other corporate communications strategies. This will take education of both corporate decision makers and customers.

Chapter 12

Cellular Services

In the past, cellular telephones enjoyed quite an exclusive standing in the public eye. They were used primarily by the nation's elite – people who could afford the finest convertibles, the most expensive foods, and telephones all over the place. Initial cellular telephone models cost well over two thousand dollars and per minute costs were outrageous.

Today, cellular providers are fighting this expensive image. Cellular phones are as commonplace as mixers and microwave ovens in many residential households and as copiers and fax machines in businesses. Monthly charges rival cable television or health club membership fees and usage rates are reasonable. Many people undoubtedly received cellular phones as Christmas presents this year – prices as low as $200.

Cellular services are a major means of access to long distance companies today. With more and more emphasis on mobile communications, customers find it easier to make long distance calls when away from the office. Sophisticated national cellular signaling networks and lighter portable phones are giving users better access to their long distance carrier's networks. With the added functionality of portable facsimile machines and voice mail services, the concept of the portable office is a reality today.

Interestingly, most of the recent growth in the cellular industry has centered on the lower end of the market, the residential or small business user. Many larger businesses are still somewhat reluctant to join the trend towards mobile personal communications. Often, they prefer older radio technologies or do not use mobile communications at all, feeling it is too expensive. As prices drop and products flourish, companies should take another look at the varied benefits that cellular communications can provide in enhancing their firm's operational efficiency.

Cellular Service Expands

Cellular service is expanding rapidly across the US. All major metropolitan areas are now outfitted and rural service areas (RSAs) are in the process of being implemented with service now.

Each potential market has been divided by the FCC into two competing licenses, a non-wireline license (called system A) and a wireline license (called system B). Depending on how a local area is defined, some metropolitan areas have more than just two carriers providing cellular service. The area in which a licensee is allowed to operate is called a Cellular Geographic Service Area (CGSA).

Cellular services are not tied to a particular telephone model. Industry standards allow for different phones to work on different cellular systems. This enhances a user's ability to switch carriers. Typically, there are three types of cellular phones:

1. **Vehicle or mobile models:** These are designed to be installed in a car. They operate off of the vehicle's electrical system. The most advanced models offer voice-activated service, allowing for truly hands-free operation by responding to voice commands.

2. **Transportable models:** Similar to the vehicle models, these are compact units with their own battery power supply, which can also receive power from other sources such as a cigarette lighter.

3. **Portable models:** These are typically smaller than the transportable phones, capable of fitting in a briefcase or pocket. They are generally less powerful as well, limiting their effective distance relative to the more powered transportable or vehicle models. Talk time can be limited to the life of the battery or the availability of a new one.

Which type of a cellular phone a company buys depends largely on the application. Many firms are realizing that cellular phones have far greater uses than originally imagined, prompting many to seriously consider the portable or transportable

Figure 12.1 Cellular telephones are getting smaller. Motorola's new MicroTAC is about the size of a wallet (photo courtesy of Motorola).

models as the preferred equipment. As batteries get smaller, the portable models sometimes become the most requested equipment.

Disaster recovery is one major reason that many firms buy transportable models. These companies have been able to survive local telecom disasters because of cellular backup facilities. By providing each major department with a cellular phone – at a fraction above the cost of local service lines – companies are assured of at least one route to and from each decision center in the case of loss of local service.

Recent disasters such as the Hinsdale, Ill. central office fire and the 1990 earthquake in San Francisco have tested such backup systems. One firm was able to maintain the flow of inbound calls despite restrictions in the local exchange carrier network, simply because it had cellular phones in the cars of its sales force. However, transportable units would have avoided the stationing of secretaries in cars in the company parking lot.

Types of Cellular Service

Most of the cellular firms offer a series of niche services, each targeted toward a specific range of uses, rather than just one bulk calling service. Among the different areas covered:

1. **Individual business plans.** These satisfy those customers that have a relatively small volume of calling on a single phone.

2. **Corporate plans.** Volume plans that group multiple phones under a single pricing and discount plan.

3. **Dispatch services.** These are designed for courier and delivery companies who have short, 0-30 second calls. They are priced to discourage longer duration calls.

4. **Residential plans.** Off-peak calling plans that offer low off-peak rates and high peak rates.

Typically, services will charge subscribers a monthly fee, plus per minute usage fees. With some services, users are given a allotment of free minutes with the monthly fee. Usage rates can vary depending on the time of day and number of phones on the account. Calls are generally rounded to the next higher full minute, but some carriers have instituted smaller increment billing, such as nearest half minute or tenth of minute. However, be sure to check how carriers bill for outbound calls – many do not have answer supervision. In these cases, the customer pays for calls from the time the number is dialed. Any money gained in shorter incremental billing could easily be lost without answer supervision.

Cellular service can result in substantial cost savings for customers in large metropolitan areas where standard telephone calls from one end of the city to another would be charged toll rates. With cellular, calls from any point in the system to another point are insensitive to distance – providing users with an opportunity to reap savings in toll charges.

How a cellular call is completed

The nation is divided into a series of "cell" groups, areas of service coverage resembling local telephone company local service areas. Each cell has a transmitter and receiver that coordinate all cellular communication in that cell. It interfaces with other company cell sites in the same way that central offices communicate in the local exchange environment – through sophisticated switch signaling.

Given the overlapping layout of the cellular networks and the local exchange carrier networks, cellular provides an excellent means of recovery in the instance of an emergency in the local serving central office. Calls may simply be routed to the serving central office of the cellular provider and then carried via the cellular network to transportable phones in company offices.

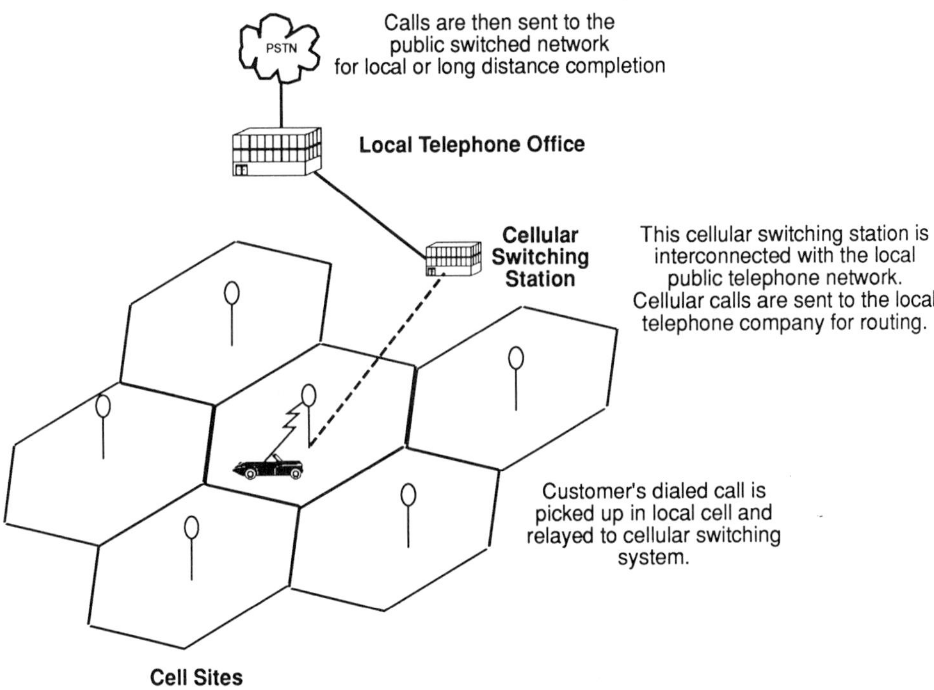

Figure 12.2 How a cellular call is completed.

Enhanced Features

One big attraction of cellular phones today is the myriad of extra features they offer – most at no or at marginal extra cost. No longer are managers just buying mobile telephone service, but mobile offices, complete with dedicated support personnel.

Look at some of the features offered by cellular providers around the nation:

- **Voice messaging.** Now a universal feature in many areas, voice mail provides users with the ability to receive messages when the cellular phone is not activated or is busy. Among the many options available, voice mail provides users with the capability to record, retrieve, send, edit, prioritize, and transfer voice messages. Users can personalize outgoing messages, passwords, even voice signatures with these voice mail systems. Cost typically ranges between $4 and $10 a month, depending on whether a user gets the basic or enhanced service.

- **Call waiting.** This is the same as its local exchange counterpart. Two calls at any particular time can be sent to a phone – any additional calls either receive a busy signal, voice mail, or are forwarded to another number.

- **Call forwarding.** This feature often includes the ability to forward all calls or just calls when the phone is busy or unanswered. What's more, these features can be used interactively. For instance, if a user has both call waiting and call forwarding-no answer, then he or she can choose not to answer the call waiting signal and let the call be forwarded to another number – which most of the time is the voice mail box.

- **Three party conferencing.** This allows users to add a third party to the call.

- **Data Services.** These provide data transmission service when coupled with a lap-top computer, fax machine, or other data transmission device.

- **Accounting Codes.** These allow users to segment their monthly bill among different clients or employees, simplifying end of the monthly cost allocation.

- **Bonus Services.** These are free services provided to customers that enhance a carrier's products and differentiate them from those of a competitor. Among the more creative ones being offered around the nation:

    1. **STAR*JAM.** Los Angeles Cellular provides this personalized traffic routing service that helps drivers avoid traffic jams, and get out of them when trapped. By dialing *JAM, the user can reach a live operator that will help the caller find a way out of any traffic mess.

    2. **Traffic Solutions.** Provided by MetroCel in Dallas, TX, this provides subscribers with traffic reports customized for the area where he or she is driving. These reports are updated every 15 minutes.

    3. **STAR*FIND.** Provided by LA Cellular, this is a "cellular yellow pages" that helps a caller find a particular business in an unfamiliar neighborhood.

    4. **Mr. Rescue.** A free emergency roadside service, which is provided by some GTE Mobilnet systems. Simply press "*HELP", 24 hours a day, 7 days a week, and a professional service technician will arrive to help within 30 minutes.

In addition, many cellular systems around the nation enable callers to look up sports scores, stock quotes, even buy amusement and entertainment tickets – all through their cellular phone. All of these provide the user with much more than basic telephone service. Cellular service is very competitive and most providers are looking for more ways to create customer loyalty for their products. Expanded feature options are the primary means by which they accomplish this.

What results when a user adds all of these features together with the latest technology is a mobile office. Users can tie telephones, fax machines, and modems into home offices, creating a communications link that increases efficiency and boosts productivity.

Calling from Outside the Local Area

Another major factor in evaluating cellular services is how different carriers deal with calls made outside the local calling area. As the nation becomes covered by cellular service areas, users traveling from one zone to the next want to have more transparency in the process.

Changing zones today is a tricky task, considering overlapping coverage areas and differing service arrangements. Moving from one cell to another within the home carrier's territory is usually no problem, outside of occasional fading and overlapping conversation at cell edges; however, changing carrier zones is more involved. Typically, a user's phone will indicate that he or she has left the home carrier calling area. This means that the user is in another carrier's territory and may have to follow any of a number of procedures to make a call.

Many neighboring carriers are starting to form "wide area" networks that are basically partnerships or handling agreements with neighboring cellular providers. These carriers are able to market their basic service to a broader coverage area in a seamless manner and lets users enjoy the advantages of a larger local cellular network. For instance, Cellular One in central New Jersey has formed alliances with Metro One to the north, Metrophone directly to the west, and Cellular One of Atlantic City, and Cellular One of Wilmington in the south, to offer a "wide area plan" that covers all of those areas at one price, without foreign billing or daily sign-on fees.

For subscribers that are not in an area with such extended service, calls may be received in one of four general manners. The most generic method is through roaming numbers. These are access numbers that allow callers trying to reach a traveling cellular user to enter the local market and attempt to secure a connection. The caller dials the roamer number, hears a tone or beep, and then enters the subscriber's cellular telephone number. If the subscriber has logged onto the local network, that call will then be completed. The caller pays for all long distance fees; the subscriber pays for all airtime and daily set-up charges.

A second manner in which calls are received outside the local service area is through "Follow Me Roaming". This "call following" service is offered by GTE Mobilnet through many cellular providers on System B throughout the country. If the home carrier offers Follow Me Roaming, then all calls will be forwarded automatically to the subscriber when he or she is on the road, following the user around the country.

When the subscriber chooses to activate Follow Me Roaming, he or she simply dials a series of numbers on the telephone keypad, telling the system to register the user. All calls are then forwarded to the subscriber – at his or her expense. Unlike other carriers, GTE Mobilnet provides its subscribers with the ability to selectively activate Follow Me Roaming in a foreign area, because users may not want to receive all calls automatically. This gives the users the option of telling few people the roam number when he or she may be located. GTE automatically deactivates Follow Me Roaming at midnight each day, or callers may do so manually at their convenience. Users must reactivate Follow Me Roaming each day when on the road. GTE's Follow Me Roaming is available in more than 300 markets in the U.S. and Canada at this time.

A third roaming system is RoamingAmerica. This is provided by Appex Cellular Information Management, Waltham, MA. Appex estimates that RoamingAmerica is available in more than 200 cities nationwide. The service is predominantly for system A customers.

RoamingAmerica provides call following similar to GTE's service, but also provides caller notification services. Caller notification allows a roamer to have incoming calls conditionally transferred to a voice announcement on the home switch. The announcement tells the calling party what city the roamer is in and provides instructions (including the long distance phone number for the foreign switch's roamer access port number) for calling the roamer on the foreign system – thereby passing the long distance charges on to the caller.

Under RoamingAmerica, subscribers automatically activate call following by placing a call from any foreign market – the user does not have to manually activate the system by pressing digits on the telephone keypad, although that option exists. Appex's system will detect this call and send an activation request to its main computer system. When the system receives this activation request, it will send a sequence of commands to both the home and roam switches to deliver the subscriber's calls to his current location. Subscribers, however, can turn off RoamingAmerica and it will automatically reset to the home voice mail features so they won't have to lose any calls or messages. GTE Follow Me Roaming subscribers would have to call long distance to their home area to reset voice mail features. Appex will also send a confirmation call to the subscriber in the foreign area noting that the phone is properly registered in the new market area. RoamingAmerica will continue forwarding calls until the caller returns to the home service area – when the phone automatically starts ringing locally – or, if the service is not deactivated, automatic cutoff occurs 24 hours following the last call made in the foreign market.

A fourth manner to receive calls is through Super Access, offered by Cellular One in several areas in California. This call forwarding service is primarily for system A customers. The system automatically registers callers when they place their first call in a foreign area, or subscribers can activate it at their own expense by dialing the appropriate digits on the phone keypad. Some Super Access customers also have caller notification. Once they make a call on a system, a message is sent to the home switch letting it know where to reach the subscriber. These instructions include the roamer access code and dialing information.

A users type of cellular service generally dictates his or her roaming service. If a user has system A at home, then he or she will likely use system A on the road. This does not necessarily have to be the case. There can be some hard economic benefits from shopping around when going traveling.

The Drawbacks of Cellular Services

Cellular telephony is by no means perfect. One often cited problem is privacy. It is possible to tap cellular communications with off-the-shelf devices, but someone would have to go to great lengths to actually make this worthwhile. As phone users move from one cell to another, conversation is switched from channel to channel, making it hard to follow. Cellular phone tapping is illegal, according to the 1986 Electronic Communications Privacy Act. Another problem is congestion, which causes degradation in conversation quality – "crosstalk" is common, leaving callers speaking to random cellular users instead of the intended party. Most firms publish a code of use in which confidential information is barred from cellular transmission.

Bell Atlantic has an option on its system in Washington D.C. that was designed to squelch any privacy problems in their territory. The carrier can outfit users with a data encryption device that scrambles communications between the mobile unit and the switching office, making eavesdropping nearly impossible. Users have to purchase a device that attaches to the car unit.

Another problem with cellular is the invasion of personal privacy. As cellular phones become an accepted way of life for senior management, it becomes hard for any particular employee to say "no" when asked to keep in touch from outside the office. Intrusion of privacy is a concern of many industrial psychologists. They suggest that executives not put employees in a situation where they carry a phone around 24 hours a day. To many, a cellular phone creates the feeling of not being able to get away from the office – a sentiment that can have a profound effect on their level of performance when actually in the office.

Evaluating Cellular Services:

When evaluating cellular services, there are some not-so-obvious factors to investigate. Ask each cellular provider about its:

1. **Rate time periods:** Are there peak and off-peak hours or day, evening, and night and weekend periods? What are the rate periods? Does one provide longer off-peak periods?

2. **Customer service:** Is there a 24-hour customer service line? Is it manned 24 hours or is someone paged after regular business hours?

3. **Call detail:** Is call detail free or is there an extra charge?

4. **Billing increments:** Are calls rounded to the next higher minute or is there 30- or 6-second billing?

5. **Answer supervision:** Are calls billed when the called party answers, or from the time the user hits the SEND key on the phone. (which can wipe out any savings due to 6-second billing)?

6. **Unanswered or busy charges:** Are there charges for unanswered or busy calls? Some carriers will charge a user for the airtime it takes to ring the phone in the user's car.

7. **Call following:** Is there automatic call following when traveling, or does someone have to find the user by dialing roaming numbers?

8. **Extended calling areas:** Has the carrier forged any alliances with neighboring carriers to form extended calling areas? These can mean lower costs for the traveller.

The Future of Cellular Services

Nevertheless, the future of cellular telephony is assured, if for no other reason than people enjoy such universal access to communications. However, technology will have to fight to keep up with soaring usage levels. Congestion is a major problem that cellular companies will have to deal with in the near future. Los Angeles is already running into problems in some areas, and expects to hit a "capacity brick wall" by early 1991.

One approach is multiplexing individual channels to increase capacity. In late 1989, Pacific Telephone Cellular and Qualcomm, Inc., San Diego, CA, previewed their Code Division Multiple Access (CDMA) digital cellular system that has the potential to provide a 20-fold increase in the capacity of available transmission spectrum. With existing cellular technology, every call is assigned a single voice channel. Under CDMA, multiple calls are spread across the same segment of spectrum through a technology similar to multiplexing. A CDMA cell has the potential to provide more than 1,200 voice channels as opposed to the 57 available in today's cells.

There are a number of side benefits besides increased capacity that will also occur with CDMA – namely, increased transmission-reception capability, fewer cell sites, better multipath resistance, and enhanced voice quality. Since the power required to transmit under CDMA is less, battery life is prolonged, giving users greater use of existing equipment. Initial tests by Pac Tel have been successful and users have noted improved transparency in call handoff from cell to cell as well.

Other technologies are on the horizon that could also significantly advance personal telephony – not just cellular but other competing or complementary technologies. The next generation of cellular phones is under development and leading cellular providers have asked the Federal Communications Commission for a market review to assess where the industry is heading and how frequency will continue to be allocated among the different technologies.

Among the more interesting futuristic products is one under development by GTE Laboratories which uses Portable Access Locations (PALs). This technology uses a compact phone unit that homes on of small, fire detector sized transceivers. These transmitting devices have smaller "footprints" than their cellular counterparts – generally the size of a large room – and therefore require less power. They are not designed for use in moving cars, but for slower mobile communications.

What is most alluring about these phones is that no one carrier controls a large area. The small coverage area allows companies to set up mini-networks for collecting and sending these calls on their own. Thus, a hotel or resort could set up a network of 200 or so detectors and issue personal phones to all registered guests, for use in the room, in the restaurant, on the golf course, or even in the hot tub. The hotel could run the calls through its own long distance network. This could be another big revenue producer for hospitality communications groups, such as the alternative operator services industry.

In the end, cellular technology will become more a part of everyday life for all of us. There has been a lot of kidding lately about the "Dick Tracy" wrist watch telephones and fax machine belt buckles. The funny thing is that these are not sounding so outlandish anymore. A typical telephone user of the near future may have one telephone number where he or she may be reached – whether on the road, in the office, on a boat, or even in an airplane. The telecommunications industry is changing rapidly, and the cellular segment is determined not to be left out.

Chapter 13

Billing

Intense competition is forcing long distance carriers to be more creative in their invoicing and management reporting for end user networks. Bills are not the same anymore.

In the early days of long distance, many monthly bills merely consisted of a cover page with key subtotals and totals. Itemized call detail was available on request, or not at all for some services, such as toll-free. Furthermore, early customer demands concentrated on whether or not the bill was correct in the first place, rather than on any abundance of network traffic and management information.

Today, the situation has changed somewhat. Customers still wonder if their monthly bill is correct. But now, the carriers have implemented sophisticated billing systems that sport a wide variety of call detail and management reporting options which provide customers with new management tools. These are available from most long distance providers.

Most carriers now consider billing to be a strategic part of their product lines. It is becoming common for billing engineers to accompany sales personnel on major account sales trips. Through enhanced reporting services, the carriers can provide end users with a level of customization that is sorely needed to help managers get a grasp on their shifting telecom costs.

From magnetic tape to computer diskettes to on-line information streams, the flow of call detail and traffic information to the end user is becoming very high-tech. Even the paper based bills are sporting the latest in computer-generated, laser-printed graphics.

Call Detail Struggles

Customers have struggled to wrestle more information from long distance providers about network traffic. The mergers and acquisitions among providers that characterized the four-year period after divestiture caused innumerable problems for the carriers' MIS (management information systems) departments who were trying to simply get a bill out. The rapid consolidation nearly paralyzed several carriers.

While AT&T's competitors tried to cope with this situation, the nation's largest carrier itself was suffering from several billing limitations. In December, 1989, AT&T began a three-year conversion process to take back billing from the local exchange carriers.

In the past, AT&T relied on the LECs for the recording, rating and printing of its low- to medium-end products. Since AT&T did not have control of the format or content of the bills, it was limited in how responsive it could be to customers – who were often being offered all sorts of enhanced options by competitors. This is still largely true today. However, the carrier has come a long way in regaining this flexibility. By the end of 1990, AT&T says it will bill for the majority of its Pro-WATS customers. By 1992, AT&T expects to be billing 80 percent of its business customers and 20 percent of its residential customers through its own facilities. Meanwhile, AT&T offers additional reporting options, such as diskette-based reporting, to customers requiring greater flexibility in monthly billing.

Competitive New Billing Services

Consolidation has not been the only bad influence on billing systems. The intense level of competition also fostered growth of carrier services which brought with them new billing challenges. For some, these new demands could be met through in-place programs; other carriers, notably US Sprint, decided to scrap the present systems and build a totally new foundation for reporting for their new services – enabling the carrier to launch billing innovations as well.

Another consequence of competition, especially among the larger accounts, is customized billing. Customer-specific billing has turned into the extra "gimme" that often wins an account away from competitors. As it turns out, this is a double-edged sword. For instance, one of the notable aspects of AT&T's Virtual Telecommunications Network Service (VTNS) Tariff 12 program is specialized billing. Many customers have opted for this high end service based on promises of an enhanced level of support – including billing services. Yet, customers have been screaming about billing errors, or worse, receiving bills six to nine months late. This is a decidedly new and problematic situation for AT&T.

Perhaps one of the reasons that call detail reports and management summaries are factoring into major account victories is that billing has taken on a strategic focus for most end user firms. No longer is it enough to get a monthly net sum of long distance calling. Managers want to know who is calling where, when, and how. They want to decipher this information to evaluate internal organizational flows and external sales channel results.

An example of the importance of call information is seen in inbound telemarketing. By watching call detail sorted by geographic areas, marketing managers are able to gauge the performance of advertising and marketing in each of those areas. Some of the more entrepreneurial companies are then matching this information with available demographical and psychological information files to get complete analyses of customer characteristics for future marketing campaigns.

The overall result is that an important shift is taking place. What many customers have bought from service bureaus and similar third-party providers in the past, they are now seeking from the carriers themselves as part of standard telephone service. Carriers are willingly taking on these tasks as they try to differentiate their products in a maturing market. The IXCs are expanding in-house service bureau-like departments to provide an extremely wide variety of information options. Additionally, the carriers earn substantial dollar savings by delivering call information in non-paper based manners – i.e., through computerized terminals, electronic mailbox delivery and various magnetic media.

Indeed, these technologies are precursors to future information flows. Electronic Data Interchange (EDI) is already a rapidly growing area of telecommunications information exchange. Many companies routinely pay employees through on-line banking transfers. It is not too outlandish to think that phone bills will be received, evaluated and paid through on-line means in the future.

Invoice and Management Information

In looking at phone bills per se, it is important to first differentiate between the two primary forms of information provided to customers each month – invoice information and management information. Invoice information covers the accounting portions of the bill – how much the user owes, where to send it, when it is due, etc. Management information provides added summaries and reports that help an end user manage the long distance services that comprise the network.

In the spectrum of end users, the low end of the market is typically characterized by only invoice and call detail information. Many of these customers are billed by the Bell operating companies. Customers do not want to spend time analyzing their bill – they just want to pay it. While there may be some level of sophistication in the low end, it is generally limited to that which allows a customer to pay for all services with one check. In these instances, the services are grouped under carrier-provided bills – but again, the emphasis is on invoicing.

At the opposite end of the market, emphasis has shifted to management reporting. It is not just knowing how much the bill is, but why. Hierarchical billing and consolidated invoicing are now considered necessary in any large corporation.

No matter where a particular firm lies in this customer spectrum, billing redesigns have made today's bills more convenient and easier to understand.

Billing Consolidation

Perhaps the two biggest billing trends in the industry are hierarchical billing and consolidated invoicing. Customer demands for increasing customization have forced the carriers to respond with flexible billing systems that can provide invoices and billing reports to any or all divisions, subdivisions, departments, etc. – all based on a customer's definitions of how his corporate accounting and reporting systems are structured.

Hierarchical billing allows each end user to define an organizational invoicing arrangement in which invoices and management reports are distributed to specified sites within the corporate organization each month. For instance, a company may decide that it wants to receive a master bill summary at the main site, division level invoices at each of its four divisions, and call detail at each of its seventeen departments. Alternatively, six of the seventeen could receive their invoices directly, with the remaining bills going to the division heads. The different permutations are truly endless.

Consolidated invoicing combines bills from individual services into one master invoice. This gets rid of the problem of multiple bills arriving at different times each month. It also enables a customer to get one unified look at all of the company's long distance charges.

MCI and US Sprint announced multilayer hierarchical and consolidated reporting systems in 1990. MCI unveiled the first version of its consolidated reporting product, called Portfolio, in late 1989, and made it commercially available in May, 1990. Portfolio is the platform upon which MCI will consolidate all of its billing and reporting information for each of its products, starting with its inbound and outbound products (except Vnet), voice grade private lines, and MCI card traffic. Four more phased releases of Portfolio are expected through the end of 1991 – Vnet and MCI Fax will be added in October, 1990 and data services will be introduced to the program in 1991. MCI introduced a location level billing option for its virtual network service, Vnet, and will extend this option to its other services in the first half of 1990.

US Sprint introduced its Invoice Processing System (IPS) in February, 1990. This hierarchical, consolidated billing system serves as Sprint's foundation for future invoicing. IPS produces two market-driven invoicing products – the FON Manager for residential and small business customers and the Management Billing Advantage (MBA) for multiproduct and multilocation business customers. FON Manager is available today; the first phase of MBA was released in second quarter, 1990. US Sprint is in the process of consolidating its various products now. The voice products are integrated under IPS; SprintNet (formerly Telenet) data products will be added in 1991, as will Sprint Gateways, Meeting Channel and Long Distance*USA services.

AT&T also is consolidating. AT&T offers the Corporate Card Execu-Bill which offers up to five levels of hierarchical billing. AT&T's Tariff 12 VTNS is fully consolidated and hierarchical, as is true of its Software Defined Network (SDN) offering. Early in 1990, the carrier started combining multiple service and location invoices for all AT&T Readyline 800, Standard 800 and AT&T WATS bills. By the end of 1990, AT&T will offer optional departmental, consolidated (cross-service) and hierarchical (or summary) billing and reporting for all inbound and outbound services.

The New Paper Bill

Present invoice and management reporting options of the major long distance carriers include a proliferation of available media – paper, magnetic tape, diskette and on-line. The creation of new services and features has created new types of billing requirements. For instance, the ability to overflow virtual network on-net calls to another number when dedicated access facilities are busy has confused the concept of terminating number call detail. Customers now need to differentiate between the number dialed and the number to which the call was eventually forwarded. As a result, bills often reflect both numbers, allowing a manager to gauge when new dedicated facilities need to be installed.

An analysis of their paper bills reveals distinctly different approaches among the major long distance carriers. AT&T began the trend towards billing overhauls in August, 1989 when it introduced several new WATS formats that included significant use of shading and boxing. US Sprint's new billing system is likewise characterized by boxing and fancy avant garde fonts. Sprint further distinguishes packaging of invoices from reporting material through the use of different color paper stock – white for invoicing and grey for management reports. The carrier takes a chance to market its services throughout the new invoices – printing a different promotional message at the bottom of each bill page. MCI's new Portfolio reports make use of bold typefaces and lines to segment and display data. The carrier accentuates its information with more than 20 different font sizes. In evaluating carrier reporting, personal tastes for graphical display must be weighed against bill utility.

Magnetic Media

Certainly, the most entrenched magnetic media is magnetic tape. All of the carriers offer some form of magnetic tape-based information. This information generally is call detail data. Only AT&T provides its actual invoice and management reports on magnetic tape.

One of the newest entrants to the field of long distance information reporting is diskette based billing. Each month, end users receive a series of diskettes that contain call detail and billing information that can be accessed, sorted, graphed, printed, and displayed in a wide variety of ways.

AT&T announced its AT&T Detail Manager (ADM) in 1988. Customers can use ADM for billing one or more telephone numbers to produce standard reports. For instance, users can create a call detail report that lists the calls in order by the city called, or an exception report that shows only those calls that lasted more than a specified number of minutes. In all, there are nine to seventeen standard reports, depending on the service, plus an unlimited number of customizable reports.

US Sprint launched its diskette product, FONView, in the summer, 1989. Like AT&T's Detail Manager, the product contains a basic program which is supplemented each month by call detail databases. FONView is more "user friendly" than its AT&T counterpart, using "pop-up" menus for instance. Among other things, the program is able to display calling by area code, the time of day, the day of the week, service, etc. The graphics interfaces have been the most popular part of the program, providing instant bar graph and line chart reports for upper management. Unlike AT&T, users can store customized report formats for monthly generation. In the summer of 1990, the carrier started converting FONView to its IPS system, laying the groundwork for broader reporting. By the end of 1990, Sprint will have hierarchical reporting available with FONView.

Telecom*USA and Cable & Wireless both offer raw call detail data on diskette – there is no formal supporting software program.

MCI is presently testing various diskette programs, one of which the carrier will launch by the end of 1990. In addition, MCI is running a number of diskette pilot programs for customers that want to use personal computers to evaluate traffic. For the most part, the carrier feels that the technology is unreliable and is looking to satisfy the same customer needs through other means.

Diskette based information is an innovation that provides many users with the first look at their network calling. Most have never had the opportunity to do any sort of traffic analysis on their monthly bills. Diskette programs allow for quick analysis and graphing of this month's traffic – not data from several months ago that somebody in the office finally got time to put into the computer. There are drawbacks, however. One of the biggest drawbacks of these diskette based systems is that they are presently non-hierarchical in nature. Each FONView diskette reflects monthly usage at the account level; each Detail Manager diskette shows calling on a main billing telephone level – neither shows cross-service, hierarchical billing information.

Another problem arises in the media itself. Diskettes are not part of the standard production process. Floppies must be produced after the invoice run, and these are often received on a delayed basis. Floppy diskettes can get damaged in the mail or in use at the office. In these instances, another set must be produced, which results in a delay in working with the information.

Yet another problem noted by carriers and customers alike are the number of diskettes required to represent a monthly bill. US Sprint limits customers to 125,000 records; at the maximum level, Sprint would need between 3 and 11 diskettes, depending on the size of diskette. AT&T does not limit the product based on number of records and will provide as many diskettes as needed to accommodate all of the customer's call records. AT&T will introduce data compression to its diskettes in the fourth quarter of 1990 to double to number of records per diskette.

On-Line Billing

Along with diskettes, interest in other electronic media, namely on-line information and electronic data interchange, is rising. On-line reporting has been a hot topic among the nation's largest users. Many are receiving direct on-line transfers through all three major carriers – some on a nearly real-time basis.

Unlike many new technological developments, applications for on-line reporting are, for the most part, just now becoming defined. This is an advance that will be truly appreciated only when new applications are developed to take advantage of the opportunity.

There are some obvious advantages today to the end users and carriers alike. Reduced cost is one – by eliminating massive printing and shipments of paper-based call detail, thousands of dollars can be saved. Also, companies can more easily maintain records of past bills, opting for storage in their own internal formats, rather than that of the carrier-provided magnetic tape. Besides these generic benefits, other advantages are likely to be application-specific. For instance, real-time call detail transfers might allow daily billing of calls – something that could greatly speed up receivables for firms that distribute or re-bill costs.

All three carriers offer some sort of on-line electronic access to traffic summaries, although each varies substantially. Most of these are reserved for the higher end services such as the virtual network and customized tariff services.

MCI's Network information Management System (NIMS) provides access to next day traffic information and near real-time access to operational and performance information. Both hourly and daily reporting are kept for the current day and the previous seven days. NIMS is part of MCI's Integrated Network Management System (INMS), MCI's platform for future enhanced network management.

AT&T offers several on-line links to its services. Its Expanded Service Management System (ESMS), part of AT&T's Unified Network Management Architecture (UNMA), allows customer to access the system's costs management elements, which include: next day on-line traffic analysis, both daily and weekly; management reports from call detail; bill or call detail on magnetic tape; and, customized decentralized billing.

One component of the ESMS which provides the customer with a series of network reports that details the cost and performance of the network is the Customer Network Administration Reports (CNAR). These reports segment usage according to domestic or international origination, and according to type of call (e.g. switched calls versus dedicated calls, on-net versus off-net, etc.). Also available are reports of long duration calls, calls made after hours, and incomplete calls on the network.

Another component of AT&T's UNMA is AT&T's Customer Traffic Data Reports (CTDR) System, which provides the customer with critical network traffic information, including daily and weekly reports on dedicated access line utilization. Daily reports are generated when preset thresholds are exceeded. Weekly reports display five days of traffic for twelve consecutive hours for each trunk group. The system stores traffic data for up to twenty-eight days for daily reports and up to three weeks for weekly reports.

Information transfer speed is always a critical issue with any on-line service. AT&T allows customers to view traffic on-line at speeds of 300 bps-9.6 kbps. Users may download 100 byte call detail records at speeds of 2.4 kbps-9.6 kbps. This is about 500 records per minute. AT&T just announced real-time access to call detail information through its various systems, which will allow any customer to receive continuous data streams from AT&T's network concerning calls as they are placed.

US Sprint offers its Integrated Network System Interface and Terminal Equipment (INSITE) system. Sprint has rolled out phase one of INSITE in 1989. Sprint cannot presently give customers billing information but plans on integrating IPS with INSITE, allowing retrieval of archived billing and network performance data by the middle of 1991. Next day on-line traffic information is available now from Sprint through INSITE. Users can get hourly data, a busy hour traffic summary for all locations or a seven-day busy hour summary. The system stores traffic data for up to 90 days for on-line access. A calculation function allows hourly traffic data to be summarized for this period.

EDI Invoicing

Telecommunications applications of EDI are just now starting to get off the ground. In general, EDI consists of frequent, high volume information transactions. Unfortunately, telecommunications exchanges of information between carriers and their corporate clients are not yet very frequent – typically, telephone invoices come once a month, a batch process. Once they do come, however, they almost always contain a very high volume of information, making EDI a perfect application for the telecommunications industry.

As a result, EDI is generally being driven by other non-telecommunications applications. Large corporations, such as food distributors, are installing EDI for in-house information dissemination applications – invoicing, purchase orders, order tracking, etc. Once all of the EDI equipment has been installed, these firms then start looking for secondary EDI applications, of which telecommunications is one.

Standards are an important part of EDI. The EDI standards for generic invoicing have become well accepted. These are known as the American National Standards Institute's (ANSI) Accredited Standards Committee (ASC) X12 810 Invoice Transaction Set standard. The current 810 standard was developed for a product environment, not the usage based telecommunications environment. As such, the carriers have had to work telecommunications invoicing information into this product-related framework. Flexibility built into the standard allows for this adaptation to telecommunications billing.

Carriers are working with customers to realize EDI transmission of telecommunications bills. The majority of customer needs for billing data is easily worked into the 810 standard; remaining invoice items can be included in the transaction via logical, mutually agreed upon arrangements between the carrier and customer. In fact, even the most detailed transmissions of itemized call detail are not limited necessarily by the 810 data format standard, but rather by the potential size of the transmissions themselves. However, a carrier and customer can often agree upon a transmission vehicle other than the typical EDI Value-Added Network (VAN) – in some cases, a direct connection (a dial-up or leased line) may be used.

The 810 standard is not optimal for all telecommunications applications. The ASC X12 Finance Subcommittee is evaluating a transaction set called 811, which has been developed by the Telecommunication Industry Forum's EDI Telephone Billing Work Group (TBWG). The TBWG is comprised of members of the BOCs, interexchange carriers, and large telecommunications customers. The 811 set would provide record formats for call detail and other presently ill-addressed billing detail information. A draft of that standard is due next year, but approval is not likely to appear until 1992 at the earliest. All the carriers are experimenting with EDI at this point despite the lack of definitive standards. Each carrier is creating its own call detail transaction definition sets until a widely-accepted standard becomes available.

AT&T was the first carrier with a formal EDI offering. AT&T transmits network services invoices using EDI. The only cost to customers receiving EDI invoices is in setting up hardware and software to receive and translate the EDI transactions, if they do not already have them in place. AT&T will charge to maintain an electronic mailbox on their network if the customer decides to use AT&T EDI. If the customer already has an existing subscription on another VAN, AT&T will deliver their 810 information to their mailbox via an internetwork gateway from AT&T EDI. Translation is supported using a customer's existing EDI software or standard off-the-shelf communications software.

Customers may request invoices on a variety of levels of detail depending on their needs. AT&T will produce invoices at a summary level to support accounts payable or in greater detail to accommodate analysis and chargeback mechanisms. AT&T currently can provide routing arrangement detail information for WATS, 800 and SDN service invoices, and circuit or circuit location detail for private line services. The level of detail may be specified for each invoice and may be changed monthly.

AT&T uses a number of ASC X12 releases of the 810 Invoice, ranging from version 2000 to 2003, depending on the particular service, customer requirements, and EDI capabilities. Every 810 transmission sent to AT&T customers is a valid ASC X12 810 that passes all needed compliance and syntax checks.

By the end of 1990, AT&T will expand EDI to include additional levels of detail and will be piloting the EDI delivery of call detail (via an X12 810) with a limited number of existing EDI billing customers. Support for VTNS EDI billing is due to begin in the third quarter of 1990. AT&T will migrate to the 811 standard when it is available and its customers request the carrier to do so. However, AT&T will not force 811 migration – it has promised to support 810 usage for as long as each customer wishes.

AT&T also accepts electronic payments for most service and equipment invoices using a variety of combinations of electronic funds transfer (EFT) and the ASC X12 820 payment order and remittance advice transaction set. Electronic payments are accepted from customers receiving either paper or electronic invoices.

MCI is running several internal trials on EDI and is in the formative stages of a complete EDI delivery program. The carrier is trying several different approaches using its own call detail transaction formats. Information is delivered by a customer-selected value-added network or MCI Mail to a mailbox for later retrieval by the end user.

US Sprint offers EDI billing presently through its SprintNet data division. The carrier will expand the offering to include voice transactions later in 1990. As with the other carriers, US Sprint will start with its own proprietary standard for call detail records and switch to 811 when it is available. The information record is modeled after the new IPS system format and covers all US Sprint services under the unified IPS system.

The Future Of Long Distance Billing

In addition to paper, tape, diskette and EDI, there are other reporting options available. Customer-unique applications are constantly arising and carriers are trying to be flexible. Many of these customer-specific options are now standard products for the carriers.

One such instance is microfiche. Microfiche is available from each of the carriers. For some customers, this is a big benefit. Often, carriers will mix billing technologies to meet customer demands. MCI, for example, is considering providing customers with hardcopy summary bills each month of only the master invoice and management reports while maintaining the call detail on-line for direct customer searches and sorts. After three-months on-line, the carrier would dump the call detail to microfiche or other storage media.

Certainly the future of long distance billing will be biased towards electronic information flows – either through magnetic media or on-line terminals – and other novel forms of media. One futuristic means that carriers may use is optical storage devices. These offer the most efficient technology for the retention of call detail information, with potential storage space measured in gigabytes of information. It will be possible to supply the customer with billing information on disks (similar to the compact disks currently used to record music) for use with their PCs, as well as using large scale optical technology to store all customer information for internal use or delivery to customers. Carriers could conceivably keep years' worth of information on just one disk.

Another more contemporary addition to billing media could be the tape cassette, either through the standard VHS video tape or some similar cassette. The problem with this sort of media is standards – generally users need to have an exact copy of the hardware and software generating the cassette in order to read and work with the information. There are literally thousands of potential combinations on the market, with none necessarily leading the way in terms of standards.

For all the carriers, consolidated billing is critical to providing future long distance services. As the market matures and margins drop, carriers will place an increasing emphasis on securing all of a customer's long distance billings. Consolidated invoices not only make the bills easy for the customer, they also make it easy for the carrier as well. Cross-service discounts and integrated service plans are easier to manage with a common billing system.

For most users, the greatest impact will be the merger of the data networks into long distance billing. For US Sprint, for instance, bringing Sprintnet under the Sprint billing umbrella makes great strides towards integrating their voice and data products. That uniformity has been lacking from all the carriers in the past and it is hard for end users to accept voice and data integration when the carriers themselves maintain separate networks and billing systems.

One thing is for certain – the emphasis on strategic billing will keep the carriers developing customer-oriented billing and reporting solutions. When combined with the trend towards greater customer control of networks, the customer will have greater access to long distance service – from ordering the service to receiving and paying the bill.

Chapter 14

Intrastate Services

A few years ago, a section on intrastate services would be packed with discrepancies and warnings for users about the phenomenal differences between intrastate and interstate offerings. Today, however, most intrastate offerings are merely extensions of their interstate counterparts. As such, the intrastate services often resemble in structure and cost their interstate siblings.

No intrastate services exist on their own – most are tied in some manner to broader interexchange offerings. The only feasible exception to this is intrastate 800 services which are often purchased without any relationship to interstate toll-free services.

As regulatory hurdles fall, more and more intrastate offerings are merged with their interstate partners, allowing users more integrated pricing and networking. At the same time, the distinctions between the two jurisdictions crumble and make way for unified product plans.

Intrastate Competition

The competitors in the intrastate market are generally the same as in the interexchange market, with notable exceptions. There are plenty of state and regional carriers that are making their niche in offering mostly intrastate network services. Demand in some of the larger states can actually maintain a number of these small competitors. California and Texas are two such states.

State services may be inter-LATA or intra-LATA in nature. A key point of contention within state utility commissions has been whether or not to allow long distance carriers to compete with the local exchange carriers for intra-LATA competition. In most areas of the country, only limited intra-LATA long distance competition is allowed.

A vast majority of states allow intra-LATA competition by resellers. Facilities based intra-LATA competition by interexchange carriers is allowed in nearly 20 states. Partial competition is allowed in certain jurisdictions. Intrastate competitors also thrive in states where resellers are allowed to compete against the local exchange carriers but facilities-based carriers cannot. This gives these resellers an edge on the larger national carriers.

This edge may not last for long. As with the national environment, the state regulatory scene is characterized by a trend towards deregulation. Basic long distance service providers are being allowed to compete side by side with local exchange counterparts, bringing potentially lower pricing and greater functionality to customer networks.

On the whole, intra-LATA competition is intensifying. Subjected to this competition and the bypass trends noted in Chapter 4, the local exchange companies must now feel as AT&T did when MCI and Sprint first started winning major accounts with competitive offerings.

Selling Intrastate Services

The intrastate environment is government by the state public utility commissions (PUCs) or public service commissions (PSCs). These bodies are authorized to grant or dismiss any intrastate carrier application, service, rate or feature that may be proposed. In general, long distance carriers that wish to do business in a particular state or district must first obtain from regulators a "certificate of public convenience and necessity." This is the carrier's authority to provide telephone service within the state boundaries. Carriers typically have to prove they have the wherewithal to provide telephone service through the appropriate financial and technical documentation. Commonly, carriers must establish subsidiaries to provide intrastate long distance service, because many states deny legal standing to out-of-state corporations. Thus, users might see names such as Sprint of Virginia or MCI of Rhode Island.

The Regulatory Status of Other Providers

Long distance carriers are not the only telecommunications bodies governed by the local PUCs. Alternative operator services (AOSs), pay phone operators, cellular providers, services aggregators, and other similar companies may be regulated by the local PUC, depending on the jurisdiction. In an era marked by declining regulation, new regulation takes a hard line stance against some of these entities. AOS firms, for instance, are having their profit margins choked as some states are limiting the rates they may charge at or below AT&T's rates. Some 900 and 976 providers are finding their businesses similarly regulated away as public utility commissions impose cost per call and calling restrictions on the dial-up information services industry.

Other firms are working hard to stave off regulation. One such group is the service aggregators. Aggregators sign up for long distance services under multi-location discount plans and get other companies to link with the plan in order to share the volume discounts attributed to the higher traffic volumes. These long distance brokers are walking a thin line between being classified by state PUCs as a "reseller" or just being a marketing agent for the carriers. Several states are reconsidering just how these entities should be regulated. Aggregators are covered in more detail in Chapter 16.

State public service commissions are getting pressure from regulated resellers and rebillers to start regulating aggregators as well. In the absence of equal regulation, these long distance companies are feeling the squeeze to convert to service aggregation themselves – something the supplier carriers definitely want to avoid.

Some service aggregators are being closely watched by PUCs for other business-related regulations. For instance, the state attorney general's office in West Virginia recently sued one aggregator, claiming the company had tried to start a "pyramid" marketing scam using service aggregation. Indeed, company brochures explained how salespersons could get a new Mercedes Benz and free trips for two to anywhere in the world by recruiting additional sales personnel.

The Status of State Deregulation

The filing requirements for carriers vary from state to state. In most instances, the PUCs require the carriers to detail all service offerings in tariffs, as they must do for the FCC at the national level. Depending on how strict the local regulations are, these tariffs may have to be submitted several weeks in advance of the requested effective date to allow for local review. Some state commissions merely "rubber stamp" the tariffs as they are filed. These states have relaxed regulation of rates, services and markets. Often, rates may be changed on extremely short notice. However, in states with strict regulation, carriers are limited to a specified annual rate of return. For each rate change, carriers must determine their cost of providing the requested service and attach supporting marginal profits justifications.

Intrastate inter-LATA competition is now practically a fact of life in the marketplace. In all of the multi-LATA states that have any appreciable amount of long distance traffic, inter-LATA intrastate competition exists.

Some of the big news in the intrastate arena still involves regulatory issues. Users look to the state boards to protect them from local telecommunications problems. Issues such as excessive alternative operator service surcharges, junk fax, and computer-controlled telemarketing have prompted numerous bills in state legislatures across the country. Similar laws have been put in place to deal with the 900 and 976 problems regarding decency and collections.

Present Intrastate Services

Intrastate services cover the same range of services as found on the interstate scale – MTS, WATS, 800, virtual networks, private lines, foreign exchange, etc. Many state services are merely carbon copies of the interstate services. The carriers simply file the same tariff on a state basis and apply Band 1 rates to all intrastate calls. This is by far the easiest approach. In some cases, however, older tariffs still exist on the local level, simply because no one in the tariff department has gotten around to changing the rates yet. Also, archaic local PUC rules sometimes keep older tariffs around because newer, lower cost tariffs are deemed threatening to the local environment. Where possible, however, there is a close match between the interstate and intrastate offerings. This facilitates the sales and billing processes.

Intrastate offerings have entered into interstate discounts plans as well. In an effort to reward customers for consolidating all of their long distance traffic, the carriers have started crediting that user's intrastate traffic to the applicable long distance discount plans. In the past, this usage did not count towards meeting usage levels.

For example, in February of 1989, AT&T announced a strategic plan to unify its intrastate and interstate tariffs that covered the commercial long distance offerings. The carrier said that this was necessary to meet the jurisdictionally unified rate structures of major competitors. The first step in this process was the inclusion of intrastate usage towards the minimum monthly usage requirements and volume discount amounts. Previously, AT&T calculated only interstate usage towards these amounts.

AT&T also started launching combined intrastate and interstate business service plans that eliminated previous cost penalties against customers who signed up for individual AT&T state plans as well as for the interstate counterparts. The resulting AT&T plans have given users a unified price covering both intra- and interstate usage. The moves were significant in the intrastate arena because they blurred the distinctions between boundaries. AT&T made their intrastate and interstate offerings more closely related than every before. Strategically, with a more unified product line, there was less opportunity for competitors to get a foot in the door to try to win business away from AT&T.

State Governments and Universities

There is a large segment of the user population that has a considerable amount of intrastate traffic. The average user company splits traffic about 50/50 between interstate and intrastate traffic. However, some institutions, such as banks and other financial houses, can have a great deal of intrastate traffic. For these companies, it pays to look closely at how the carriers rate intrastate calls.

Of particular interest is how intra-LATA calls are completed, as mentioned in Chapter 2. For the intra-LATA calls, customers should verify if a major long distance carrier's rates are less than the intrastate exchange carrier's rates. The long distance carriers typically do not make up any negative differentials for customers unless asked, and usually they only refund money for their larger customers.

Sometimes the local and regional carriers will give users better deals on intrastate calls than the national long distance carriers, because they may have special pricing plans just for that area. They may also be one of the few entities allowed to carry the intra-LATA traffic.

One of the hotly contested segments of the long distance market is state governments and universities. Each of the interexchange carriers have been offering extremely price competitive offerings designed to win these lucrative and prestigious accounts. Several of the carriers went so far as to design special programs for these customers. AT&T's Tariff 16 is a state network document in which AT&T files for special state pricing. US Sprint has a similar program called SIGN (State Institutions and Government Networks).

The governmental market became very visible with the awarding of the federal government's FTS-2000 contract. That contract, won by AT&T and Sprint, was for a ten-year multibillion dollar network. Already, the contracts added to the initial contract have made the effort quite profitable.

The Future of Intrastate Services

In all, intrastate services are a lot less of a problem for managers than they were several years ago. Where there used to be a number of separately priced long distance offerings, most interexchange offerings today are functions of or options off of interstate services. As such, there are fewer and fewer surprises.

With deregulation continuing as planned, there is little reason to believe that the future outlook will be any different. State regulation is emphasizing the enhanced services arena in which users get bilked or swindled. They are largely leaving basic long distance services alone which they consider more mature products, subject to a highly competitive marketplace.

Chapter 15

International Services

International services are a vital area for growth for each of the long distance carriers. U.S. firms are also increasingly searching for growth opportunities overseas. This chapter primarily considers international services from the viewpoint of U.S.-based carriers and U.S. firms doing business overseas. The focus is predominantly on Western Europe, particularly the U.K., largely due to the importance of that region to U.S.-based firms. Other regions of the world, namely the Pacific Rim, are likewise important but are not discussed here.

The importance of the global market was emphasized in early 1989 with the awarding of the first large contracts for "global networks." While there have been international private line networks in existence for years, only now are packages being developed which explicitly address the concerns and requirements of companies with international networks.

For the end user, trying to negotiate the installation of an international network is certainly not easy. Issues such as language, different time zones, varying standards, restrictions on data and fax transmission, and currency transfers are some of the problems to be confronted. United States users wishing to install a global network will spend a great deal of time deciding where nodes can be placed. Lack of capacity may limit the number of possible countries, as would lack of complementary access options. Yet international services reflect a logical expansion from U.S. foundation networks. Networks will become easier to install as long-standing barriers to market entry are removed, particularly in Western Europe.

The European Landscape

In 1992, the European Community will start removing administrative and operational barriers among its members. Telecommunications is but one of the many areas encompassed by the economic unification of Western Europe.

The changes will have a massive effect on users in the U.S. and abroad. Within Europe, companies will be able to integrate previously distinct corporate divisions and subsidiaries, allowing for more efficient day-to-day management. Communication will play a big role in supporting management. Barriers to information flow and administrative constraints have prevented such companies from being integrated in the past.

For most of this century, governments have kept outside companies from competing against the national telephone monopolies to protect their investment. The privitization of British Telecom in 1983 was a case study for the benefits that can accrue from nongovernmental ownership. The desire for the newest technology is making many countries re-examine their priorities. The lack of a strong telecommunication infrastructure is hurting some national economies. This is forcing many governments to turn to outside experts for help.

France and West Germany have been actively seeking outside expertise. Cellular, paging, beeper, and cable television licenses have been put up for bid by anyone with the technical know-how to install the systems.

The political reforms in Eastern Europe have opened the path for similar opportunity. Those nations need complete overhauls of their local and long distance systems, and are willing to share future profits to do so.

Despite the overall trend towards privatization and liberalization (and the softening of trade barriers), many impediments remain. Few countries allow private companies to provide POTS and most ventures with firms from overseas are accomplished through partnerships.

International fiber optic projects continue to be developed to handle the increasing flow of traffic among the world's nations. Commitments for the tenth Transatlantic Telecommunications (TAT-10) cable are already well underway, just as TAT-9 is becoming ready for service. The capacity of these cables cannot be understated. TAT-9, for instance, which links the U.S. and Canada with France, the U.K., and Spain, can carry more than 80,000 simultaneous voice conversations. TAT-9 cost more than $400 million to complete and will last an estimated 25 years. New cable systems in the Caribbean and Pacific Basin also are adding to the global availability of fiber bandwidth.

International satellite facilities are also quite abundant. Satellite has served as the main mode of transport for video images. To date, satellite has been the main backup for fiber in case of an outage. However, the new fiber cables will be able to provide near instantaneous backup for each other in the event of an emergency, possibly bypassing the traditional satellite option.

International Standards Remain Important

One area that interests both carriers and users alike is standards. When a user tries to interconnect its U.S. domestic network with locations overseas, there may be problems, unless the U.S. carrier provides some interface for the customer. European standards for transmission speed and equipment type are not generally directly supported in the U.S. long distance network.

For years, the U.S. has standardized on T-1, for instance, as the major core backbone circuit. This is a speed of 1.544 Mbps. In Europe, however, the carriers have standardized on E-1, which is thirty 64 kbps channels, or 2.048 Mbps. Thus, for a U.S.-based customer trying to hook a high capacity route into a West Berlin office, there may not be a T-1 termination capability.

A similar problem exists when a 64 kbps terminal in the U.K. wants to communicate with a 56 kbps terminal in the U.S. That 64 kbps terminal has to scale down its transmission speed to 56 kbps to communicate. In fact, most users recommend maintaining one speed or another in a network to avoid confusion. The only common speed of all the terminals is 56 kbps.

The carriers are only starting to offer a T-1/E-1 conversion capability. Switched 64 kbps services are also just beginning, mostly as part of ISDN services. Initial demand for these services is likely to be high in New York and San Francisco where a vast majority of U.S.-based international business resides.

ISDN promises to help bridge the U.S. domestic services with the international standards. With 64 kbps as the foundation, ISDN-based services represent a global standard for transmission. The key will be the availability of compatible access in the U.S. local exchange network, as mentioned in Chapter 3.

In general, U.S. firms are attracted to the U.K., France, the Netherlands, and Japan because these countries tend to have the most advanced networks and maintain sales offices in the U.S., making it easier to plan a network. Also, a number of these countries support some of the U.S standards that are not compatible with other international standards.

The competition between countries for U.S. corporate overseas office sites is keen. Typically at stake is not only the placement of an overseas office, but a launching site for future expansion as well. Indeed, many companies use overseas offices as transit sites – traffic is aggregated to that one site for dispersal to other points throughout the continent.

The U.K. and France often find themselves bidding against one another for corporate sites. The wealth of fiber capacity to each country makes both ideal for linking U.S. sites to offices located in either or both countries.

Operating Agreements

International services are one of the last of AT&T's monopolies. For years, AT&T has been the major provider of international voice and data service. Practically all of its domestic services have corresponding counterparts in the international arena – Accunet has International Accunet, MTS has International MTS, etc. AT&T's present coverage is truly worldwide. The carrier has the necessary agreements in place to provide communications between the U.S. and the world, with few exceptions. Among those few not served are Vietnam, Cambodia, North Korea, and Albania.

As MCI, US Sprint, and others advance on AT&T's domestic market, its international market is relatively well-protected because of the massive effort required to negotiate international agreements for the long distance traffic. Certainly, to duplicate those efforts will take years.

Operating agreements allow a carrier to connect directly to a national network, based upon established rules of interconnection. Traffic carried between the two countries is subject to a group of revenue-sharing agreements, known as "accounting rates." Accounting rates represent the basis for division of revenues between the national telephone company, whether privately owned or the PTT, and the U.S. carrier. In general, accounting rates are split evenly. Thus, if an accounting rate is $1.60, the national telephone company and the U.S. company each would receive $.80.

Accounting rates are set in three different monetary forms – U.S. dollars, gold francs, and special drawing rights (SDR). The U.S. Dollar is the most common form of calculating accounting rates. An SDR is a derived rate, compiled from an average of five countries' currencies. As of August 1990, an SDR is valued at approximately $1.26 U.S. Dollars.

Carriers are required to use the same accounting rates for similar services to the same countries. AT&T cannot file a different rate than MCI or US Sprint, unless the company can prove that it has a unique service not offered by another carrier.

Collection Rates

Accounting rates are relatively straightforward. What confuses the issue is the tariff (collection) rates – the rate charged to users. Collection rates are becoming more competitive as the international marketplace heats up. However, the margins for competitive bidding are somewhat more restricted than in the domestic marketplace. First, the U.S. carrier only gets half of the revenues. The national phone company receives the same amount of compensation for the completion of a call regardless of which carrier is handling it on the U.S. end. The U.S. carrier may decide to take less – by charging the customer less.

Consider the $1.60 accounting rate as an example. If the U.S. company is "collecting" $1.45 per minute of calling, the "accounting" split between the companies is $.80 for the national carrier, $.65 for the U.S. carrier. In some cases, the U.S. carriers may actually generate negative profit, in order to offer highly competitive rates to U.S. consumers and to get other more profitable portions of their business.

In general, since the introduction of competition into the U.S. telecommunications market, competitive forces have pushed down the collection rates of U.S. carrier services. Accounting rates, however, have not kept pace. They have only recently started dropping and still at a much slower rate.

The Global Political Picture

For the U.S., accounting and collection rates contribute to the U.S. trade deficit, and are regarded as a serious political problem. The ratio of revenue collected by U.S. carriers compared to the payout of accounting rate settlements has been steadily falling. This loss in profits for the carriers has been countered by a steady increase of international traffic. Between 1975 and 1988, international traffic growth increased from 14.2 percent to 37.4 percent.

However, this growth has not been equally divided between originating and terminating traffic – the majority of growth in international calling has occurred in outbound traffic, stimulated by lower prices in the U.S. market. The result is that U.S. carriers making payments for settlement of international accounting rates in nearly all cases.

In July, 1990, the Federal Communications Commission said that it was targeting these accounting rates for revision to help reduce the net outflow of telecommunications payments. The action is expected to place a great deal of pressure on foreign governments to renegotiate more balanced rates.

Growth-Based Accounting Rates

To generate greater inbound traffic to the U.S., AT&T and other carriers have introduced growth-based accounting rates. In such cases, accounting rates are adjusted after a certain level of minutes is reached. This threshold level may be a fixed amount, a relative imbalance amount, or simply all imbalance traffic.

For example, the growth rate for Brazil is $1.20 U.S., which is reduced from the normal $2.50 U.S. rate. The growth rate is in effect after 38,800,000 minutes of calling between the U.S. and Brazil.

The U.S. hopes that growth-based rates will encourage national administrations to stimulate calling into the U.S. and help bring down the U.S. trade deficit. In the end, growth-based accounting will likely mean lower rates for users.

Competition in the International Arena

Initial competition to AT&T has mirrored the early days of long distance, concentrating on the most basic services. International MTS has been the first targeted product. Most of AT&T's competitors are offering comparable international direct dial services, often at the same rates as AT&T's. In reality, many of these carriers are merely purchasing AT&T IMTS through its high-volume international voice service and reselling it to their own customers.

Meanwhile, however, these competitors are aggressively pursuing their own international agreements. As these are signed, the carriers are able to initiate their own service – not resold AT&T service. By using their own facilities, the competitors will have a greater level of pricing flexibility with which to win accounts. As of the middle of 1990, both US Sprint and MCI had negotiated major traffic agreements with those countries to which most of the U.S.-originated traffic terminates.

International Networking

After signing agreements, competitors have to install international networks. There is one big difference between the U.S. domestic networks and the international networks – U.S. carriers do not own facilities in other countries. In completing a switched voice call, the U.S. carrier will concentrate its traffic at international gateway points, typically New York or San Francisco, and send the call over its international facilities to the national administration's gateway. From there, the national carrier completes the call over its domestic network.

In building international networks to carry traffic, the U.S. long distance carriers are effectively acting as large users – they buy high-volume facilities from the overseas carriers over which they route customer traffic.

This has special implications for advanced network services, such as virtual and toll-free services, because most creative intelligent network services involve the use of the national entity's switching intelligence. An advanced service merely links the capabilities of the diverse partners through intelligent network interfaces.

When a carrier initiates service to a national country, the call actually may be completed in any of a number of ways. The call could be sent directly, as described above, or it could be sent through another country acting as a hub. Such countries are called "transit" countries. Alternatively, the carrier could be simply reselling another carrier's service. In that case, the U.S. carrier routes the call to the nearest international gateway and hands it off to another U.S. carrier who has operating agreements, and, in turn, passes the call to the national telephone company.

Networking on an international scale is, in some ways, similar to that in the U.S. Carriers try to locate their traffic hubs in strategic transit countries to cut costs. Regional calls are sent over high-capacity circuits to these hubs for completion to neighboring countries. Belgium, for instance, serves as a key branching point for traffic to northern Europe and Scandinavia.

Marketing drives the networking decision as well. Carriers determine how much traffic is going to each country. High-volume countries such as the U.K. and Japan are the most attractive host countries. In building their international networks, AT&T's competitors singled out the most called countries as the first installed, not just because of the higher profits but also because the network can be more efficiently used with the higher traffic volumes.

Politics cannot be left out of the international networking decision. Often, the decision to enter into a traffic agreement with a particular country is a function of the ease with which an agreement may be obtained. The Netherlands, for instance, has promoted liberalized market entry for overseas firms in its telecommunications policies, prompting many carriers to install service to that country.

International Construction

As with the domestic interexchange business, it is quite costly to build international communications facilities, whether they are undersea fiber optic cables or geostationary satellites. These high costs force carriers to band together to construct the transmission facilities. Thus, on any particular project, one might find AT&T, MCI, US Sprint, British Telecom (BT), France Telecom, Cable & Wireless, and others joining forces to fund construction.

In general, there are two forms of international communications facilities – satellite and cable. All of the recent cable construction has been fiber optic based, largely due to the vast capacity allowed by that technology.

There is an ongoing contest between the satellite and cable forces for the larger share of international communications. Video transmission, and a large portion of data as well, has always been sent over satellite facilities. However, the half-second delay caused by sending signals up to the satellite and back has made cable the preferred mode for voice telecommunications.

With the vast capacity available in fiber optic cabling, data and video transmission are now quite viable through cable. Consequently, price competition is starting to develop as fiber and satellite based transmission carriers vie for contracts.

Mergers, Acquisitions, Partnerships

With the partnerships in construction come other partnerships as well. Mergers, acquisitions, affiliations, partnerships, and other business arrangements are starting to take place that will shape the global competitive marketplace in the future.

BT bought the international value-added network provider Tymnet. AT&T invested in the Istel. MCI bought into Infonet. Sprint has a marketing alliance with Cable & Wireless. Each of the major players are positioning themselves for the upcoming surge in international networking that will result from the lifting of trade barriers in Western Europe in 1992.

There are challenges to doing business and maintaining partnerships overseas, requiring sensitivity to cultural differences as well as knowledge of the technical and physical (distance and time) realities.

The BOCs Go Global

The Bell operating companies have been scrambling for a piece of the pie in Europe as well. Restrictions in the U.S. have pushed the BOCs to compete for lucrative contracts overseas, and they are winning them. US West is looking to build the world's longest fiber optic cable across the Soviet Union and will build a cellular phone system in Hungary, the first in Eastern Europe. A consortium that includes Pacific Telesis will build the first private cellular system in West Germany (beating bids from five other BOCs).

The telecommunications market in Western Europe is becoming larger and more competitive and less regulated, too. The European market is seemingly endless for these BOCs; the mobile communications market alone can exceed $10 billion by 1992.

The overseas markets are natural extensions for the BOCs, whose domestic holdings include cellular, paging, computer maintenance, financial services, office equipment, and other areas. The BOCs are anxious to get hands-on experience in areas that they are precluded from entering domestically – long distance, electronic information services, cable television, and equipment manufacturing. The overseas markets allow them the freedom from domestic regulation to explore these areas.

International Virtual Network Services

Many of the international services are merely extensions of previously discussed domestic counterparts. These include international MTS, WATS, 800, 900, facsimile networks, private line services, and travel cards. Attention has focused initially on the U.K. as this is one of the largest communications markets in the world. AT&T, MCI, US Sprint, and others have targeted this country as the first major battlefield for international services. All three of the major carriers have their own agreements in place with the two British carriers, British Telecom Ltd. and Mercury Communications. The three U.S. carriers have each launched special pricing packages for calling to the U.K., and prices are dropping.

While most international long distance services have been in place for a while, most users are awaiting the introduction of global virtual network service. This will be the first intelligent service to take advantage of the new partnerships in the global marketplace.

Extending virtual network capabilities and network intelligence to the U.K. may not be an easy task. One of the biggest problems is the basic difference between the U.S. and U.K. telephone systems. There is no equal access in the U.K., for instance. British Telecom is the local and long distance carrier for a vast majority of the population. Thus, each local exchange is a point of presence for long distance as well, just like the old Bell system in the U.S.

To subscribe to Mercury, the customer must get a new telephone line installed (if the user is in one of Mercury's serving cities) or have an autodialer pulse a series of digits before each call. Under this latter method, BT actually carries the call through the network to a Mercury point of presence.

In addition, there is a variable length dialing plan in the U.K. The North American Dialing Plan, with its area codes and ten-digit numbers, does not apply there. Instead, calls can be seven, eight, or nine digits in length, creating problems for U.S. number translation software.

These and other differences must be accommodated for interconnection to occur.

Virtual Networks and British Telecom

British Telecom is looking at developing a virtual network service, called "Feature Net," but more for interconnection with the U.S. carriers than as a stand-alone product for the U.K. BT questions the market for such a product in the U.K. – users have never adopted the advanced forms of long distance service as have U.S. users.

BT has been taking a cautious approach towards linking with international carriers, particularly the U.S. ones. The company is the major carrier in the U.K. and believes itself compelled to treat all overseas carriers alike – that is, not favor one or another in the way it sets up international network connections. In the same vein, BT wants to ensure that it does not lock itself into any one carrier's proprietary technology, thereby precluding interconnection with other global carriers.

The design of the network interface protocols between foreign and domestic U.S. products is a critical issue for U.S. carriers. Unless the interconnection protocols follow the closed user group standards, carriers would be left without any guiding standard for interconnection. There is no international standard for virtual network interconnections. Rather, ad hoc methodologies that have been used over time are in place. A carrier would prefer to see the country's carrier install an interconnection standard that favors its virtual network product, or, at the least, install one that is ubiquitous and does not favor a competitor.

British Telecom, being the majority carrier in the U.K., shoulders the responsibility to remain open to connection with any overseas carrier that requests it. In designing a new service, BT has to decide which vendor's equipment to use and how it will mesh with other carrier networks around the world. British Telecom has most of its domestic network intelligence residing in a subnetwork of nine AT&T 5ESS switches. All of its 800 and 900 services are driven by software programs loaded onto this subnetwork. BT also has a local network comprised of GEC Plessey System X and Ericsson AXE-10 switches. While the 5ESS network contains most of the advanced routing and screening functionality of the BT system, the carrier is looking to put more intelligence into the local network to provide greater flexibility to create new services.

Interlinking Global Networks

As with domestic services, signaling is important, particularly if one is talking about linking the highest level of long distance services available today. BT's network operates on a signaling protocol similar to SS7, called "BTC7." Linking the BTC7 signaling network with overseas SS7 networks will take some time, for many of the same reasons that BOC SS7 interconnection will be slow – lack of standards. SS7 has been defined for the very basic interconnections, but detailed connection for highly advanced services will require more stringent standards. Lacking standards, few carriers want to endure the trouble of creating specific technical interfaces with an overseas carrier if they are to change in a year or two.

Meanwhile, overseas carriers have to settle for simply linking with the British Telecom network. The most sophisticated connection that BT is offering non-U.K. carriers for their virtual network products is the City Direct Service. This allows customers in the U.K. to link with U.S. private or virtual networks via dedicated connections. For virtual networks, this is effectively a dedicated-to-dedicated network call.

AT&T is now making use of this service, providing the means for a customer to have an SDN site in the U.K. AT&T sees this as one of the first steps towards its developing "Global SDN" concept. AT&T calls its half of the service, International City Center Service (ICCS).

An SDN customer installs a dedicated link in the U.S. to the nearest AT&T SDN central office. Calls are switched to the international gateway on the East Coast of the U.S. for switching over international facilities to an Ericsson AXE-10 switch in the U.K. From there, the call is switched through the BT network to terminate at a dedicated facility at the customer's premises.

In a sense, ICCS is simply abbreviated dialing with advanced SDN functionality. The ICCS service has a seven-digit on-net international dialing plan just like other on-net numbers. Interestingly, the call must undergo two translations – once in the U.S. when the call is converted from the dialed seven-digit number to a special customer-identifying seven-digit number, and again on the U.K. end when the seven-digit number is translated into the customer's on-net location number. BT performs this double translation to prevent being locked into AT&T's dialing plans.

British Telecom's Future Virtual Network Service

The next step for BT after City Direct will be a full-blown virtual network service, called Feature Net. BT has announced that it will interconnect with MCI by the end of 1990 to start offering virtual network access and egress in the local market in the U.K., but initial features will be limited.

British Telecom is working to enhance the intelligence in its network for products for which it has a demand today. Both 800 and 900 products fall into that category. The company began offering its first ISDN service, automatic number identification, at the end of 1989.

However, virtual network services will be slow in coming. The near-term requirements for virtual network services in the U.K. are international – driven by the U.S. carrier demand for intelligent interconnection.

Other advanced products are being developed, based on market demand. Market demand in this case refers to both U.K. business needs, as well as overseas carrier requirements.

BT has been skeptical about the U.K. users' demand since AT&T was unable to muster any customers for its ICCS service in the first several months of its existence. Nearly a year after introduction, AT&T acquired its first ICCS clients.

Implementing a virtual network in the U.K. will not be easy. BT will have to tackle the issue of switched access. The core differences between the U.S. and U.K. telephone systems truly come to light in trying to provide switched access into an international virtual network node. In the U.S., a customer merely presubscribes his line to the virtual network carrier and all calls are forwarded to that carrier's POP. The carrier then screens the ANI for routing information and sends the call on its way. BT does not have a counterpart system in the U.K.

Virtual Networks and Mercury Communications

Mercury Communications has an advantage in being the minority carrier. The smaller competitor in the U.K. by a large margin stands to make great strides forward by teaming up directly with one of the U.S. carriers.

Mercury has been adopting an MCI or US Sprint approach to the market in the U.K. The carrier has been targeting the largest customers in 75 major metropolitan areas for its initial services. These customers are the most profitable and will give Mercury the funds needed to grow until it can compete against BT countrywide.

Right now, Mercury is the "little guy," however. Market share is around 10 percent of the business market in the U.K. and the carrier is looking to use its network as an advantage. Mercury has been putting a lot of emphasis on its fiber optic network, reminiscent of US Sprint's marketing theme for the past few years.

Another advantage is its international connections to the U.S. Cable & Wireless, parent company of Mercury, owns the U.K. end of the Private Transatlantic Telecommunications Systems Inc. (PTAT) cable. PTAT is a privately-funded fiber optic cable stretching between the U.S. and the U.K., capable of carrying more than 85,000 simultaneous conversations. The recent acquisition of the U.S. end of the PTAT cable by US Sprint could provide the edge for which Mercury is looking in international traffic.

When the acquisition was announced, both US Sprint and Cable & Wireless announced that they intended to work together to grow globally. They announced an international marketing alliance, called GLOBAL FON (Fiber Optic Network), to develop products jointly, beginning with an international virtual private network service (IVPN) and an international calling card. These services would be primarily based on US Sprint's VPN and FONCARD services, respectively. In addition, private line and other services would be introduced as the market and technology demanded.

This marriage has the potential of being far more advantageous for the two companies than most others in the industry will admit. Compatibility abounds. US Sprint and Cable & Wireless believe that fiber optic networks are the building blocks of the future. Both have Northern switches (DMS-250 and DMS-300) in the network, and are looking to move ahead quickly in the market.

Exporting U.S. Intelligence Abroad

The key factor is where the intelligence resides. With the present AT&T and MCI products, there is no sophisticated virtual network processing intelligence stored overseas. Under BT's City Direct services with AT&T, the intelligence resides largely in AT&T's network in the U.S. BT is simply shipping a call to a predefined station in the U.S.

The same is true of MCI's Vnet I-DAL service. The customer is getting a link into a domestic Vnet, and that link will have all the intelligence afforded a domestic circuit, but the routing and switching information is still resident in the U.S.

Under the Mercury-US Sprint agreement, however, the intelligence actually resides in the U.K. switches, giving both Sprint and Mercury customers extended call processing capabilities from the U.K. If a user dials a Paris number from London, that call will go directly to Paris from London, based on routing information given to it within the U.K., not the U.S.

The installation of Sprint's programming in Mercury's network could realistically cut years off Mercury's network development schedule. With access to Cable &

Wireless's domestic network, Sprint could very quickly gain a foothold, through Mercury, in the British market, easily leapfrogging AT&T and MCI in the process.

Cable & Wireless's reach does not cover only the U.K. The carrier has extensive holdings throughout the world, including the Caribbean, the Pacific Basin, the Middle East, the Indian Ocean region, Africa, and North America. The carrier also holds an interest in the North Pacific Cable (NPC), the construction of which is to begin in 1990.

Future versions of the AT&T and MCI overseas virtual network products will have the intelligence resident in switches overseas, giving the global network integrated intelligence for switching calls.

Access Burden on Customer

Local access to long distance networks will be a problem for all carriers, however. While the U.S. carriers will be making their virtual network products available internationally, access is largely the responsibility of the overseas office of the U.S. firm. Consider the ICCS/City Direct example. Access to AT&T's SDN product is from the AXE international gateway switch. In order to gain any sort of interconnection with AT&T's intelligence, the customer must run a dedicated facility from wherever its site is located, however far away.

Until BT installs advanced signaling systems and intelligent software interfaces in its network, customers will have to send their calls through AT&T for real SDN functionality. The same is true of MCI and US Sprint as well, although US Sprint may get some relief with the Mercury connection, as already stated.

As for switched access, the U.S. carriers will have to look at some sort of toll-free access with authorization codes. This would work similarly to travel card options – dial a toll-free number (not an 800 number, remember) and input the desired number plus numerous digits for validation. The problem with this is the security issue with which so many carriers have had to deal in the U.S.

Toll-free virtual network access would help all of the carriers, and it should not be too long until it is available throughout the U.K. BT is working with all of the carriers to use the intelligence of its 5ESS network to drive some remote network access options. Until then, however, access will be somewhat burdensome on the international customer.

Travel cards can also raise interesting international access issues. With the 800 type of access for travel cards, many countries have network tones that are different from those in the U.S. This creates problems in accepting and validating authorization code entries; in some countries, all the phones are rotary.

International Virtual Net Expansion Plans

International virtual network expansion plans by AT&T, MCI, and US Sprint are aggressive. Customers have told the carriers that they need international calling and pricing from their domestic virtual network products. The locations that users have requested are roughly the 15 largest countries in terms of revenues.

There are several factors that determine whether a carrier will expand to a particular country. The national telecommunications entity must be willing to negotiate in the first place. This is often a factor of the political and economic climate in the country.

Expansion is also determined by the level of competition, technology, geography, and market needs of that particular country. The fact that Mercury Communications is working with US Sprint to develop a complete domestic virtual product is putting a lot of pressure on BT to introduce a similar service. Also, there may not be the level of technology available in a given country necessary to link virtual services. Further, the country might not need a geographically ubiquitous service; rather, a network between major cities might suffice. Finally, some countries are more interested in certain features than others. The French, for instance, are looking more for switched data and closed user group related functions. These demands may or may not play upon the capabilities of any one carrier's network.

AT&T

AT&T has already completed its expansion plans for global SDN. Outside of the British Telecom ICCS connections, there is a big void over the next year or two until BT implements its national virtual network. This will give overseas carriers enhanced access and egress capabilities, as well as at least the most basic virtual network features.

After the U.K., AT&T is introducing SDN ICCS access to Australia, Japan and the Netherlands as the first step of launching virtual network capabilities to those countries. The Netherlands is one of the more progressive European countries, from a telecommunications standpoint, and is making it easy for overseas carriers to link with the national systems. Also, there are a number of multinational firms headquartered in the Netherlands, many with interests in the U.K. Australia and Japan also rank as two of the higher traffic countries.

The manner in which AT&T will interconnect with these different countries varies substantially.

The Netherlands will initially provide only special access capabilities to SDN customers. As with the U.K., there is a great deal of uncertainty over the domestic need for a virtual network product. It is possible that instead of a full-blown nationwide virtual network service, the Netherlands may opt for a virtual product that just covers major metropolitan areas.

Japan and Australia are entirely different cases. Both countries have committed themselves to developing domestic virtual private network services, complete with multiple access arrangements – switched or dedicated – as well as a subset of features.

Both countries will be pursuing these virtual network goals in a series of phases. The initial phase for both will be much more robust than ICCS-City Direct offering to the U.K. It will likely be characterized by multiple-city access points and possibly multiple access methods. Japan and Australia will at least feature special access connections, but switched access is being evaluated.

Later phases will be feature-rich, incorporating both overseas carriers' feature requirements as well as particular domestic needs. Geographic coverage will expand.

MCI

MCI will continue to compete head-to-head with AT&T, and this means following the carrier overseas as well. MCI introduced Vnet overseas in 1990, first through its I-DAL service, and then through the BT Feature Net connection. MCI will precede AT&T to the U.K. with this offering. Japan will soon follow.

Still, in many cases, MCI will be following AT&T's lead. Unlike the U.K., where BT has taken measures to avoid an advantage for any one overseas carrier, market entry by AT&T may be more welcome elsewhere. With AT&T paving the way, to some degree, it is easy to follow afterwards and install the same service. However, for reasons mentioned earlier, this is dangerous because it gives AT&T a chance to install interfaces that are preferential to its network.

MCI will try to expand as many of its features as possible overseas, as soon as possible. Since the carrier is emphasizing its vast number of features in domestic sales presentations, one can expect MCI to promote itself in international features as well.

US Sprint

US Sprint will largely follow AT&T's global approach as it makes VPN available to international customers.

Sprint's product introduction will mirror much of its U.S. expansion. First, the company will try to emulate the functionality provided under existing international private networks. Part of this is interfacing with the PBX and its features, and part is in dial plan emulation. Then, Sprint will launch the more sophisticated features.

As with AT&T and MCI, enhanced features will be hard for Sprint to launch initially, at least without the help of a national telecommunications administration. Sprint has the best opportunity to move forward quickly with those countries where carriers have the same switch – Northern DMS-250, for example, in the U.K., Canada, and Ireland.

The technical interfaces with other countries will be straightforward for Sprint, but the carrier may have to rely on ISDN closed user group functionality to bring the more sophisticated features to the market. Sprint started providing ISDN-related features internationally as offshoots of their domestic counterparts in 1990, but initial features will rely on internal signaling schemes. Full conformity with ISDN standards will occur in early 1991.

International Networking Issues

International networking is quite different from domestic networking because of the vast number of different players that may be involved. While a user might create a domestic network linking ten cities, chances are that no more than two long distance carriers would coordinate that network for the customer. On the international scene, depending on the carrier's relationships with the national PTTs, the customer might find itself negotiating each piece part of the network with all of the vendors involved.

This is where an overseas sales presence becomes important. More and more countries are opening sales offices in the U.S., primarily in New York, to help educate American telecom managers and to design possible network solutions. One of the biggest problems that these non-U.S. carriers face is a largely uninformed buying public. Once the carriers go beyond IDDD and private lines, most customers get lost.

Aside from selling international services, these carriers want to encourage users to buy their domestic products as well, however. France Telecom, for instance, would very much like to sell each international corporate user a Minitel network – but most U.S. users are unfamiliar with this videotext box. The PTTs also want to encourage companies to be innovative in how they use the core telecommunications networks at home – this is where digital local networks assume a new importance.

A big trend in the U.S. international marketplace is "one-stop shopping" – the ability of a customer to go to one carrier and purchase all of its international network needs. Thus, a customer could go to AT&T and order circuits in the U.S., the U.K., France and Japan from one office. Moreover, the customer would receive one bill each month, not four.

Each of the U.S. carriers is putting a lot of effort into making it easy for customers to order all their services from one point of contact. One-stop shopping deals between carriers and countries are being announced regularly.

Global virtual networks also raise an interesting question about billing capabilities of the carriers. Most people have been billed two or three months later for calling card calls because of delays in processing by the BOCs. This problem can be magnified when talking about international billing records. Depending on how this issue is addressed in negotiations, users may have problems squaring monthly bills with actual usage.

The Future of International Services

In the future, international services will continue to grow. As U.S. product and labor costs continue to outpace the world market, many manufacturers will look overseas to cut production costs. Locating manufacturing facilities overseas will not be as daunting if top-level management stays in close contact via sophisticated telecommunications facilities.

This is not as far away as many users might think. The national governments are realizing that it is in their interest to have a strong telecommunications infrastructure to support the domestic economy. The telecommunications infrastructure is the highway toward economic modernity in less developed countries and continued growth in developed countries. An infrastructure with viable links to international networks, service, and carriers is also a path to further worldwide integration and interdependence.

Chapter 16

Carrier Pricing Strategies

In the competitive long distance market, the carriers are all struggling to win customers, continued growth and long-term success. In most instances, as we have seen in the preceding fifteen chapters, pricing is an integral component of carrier strategies to capture a valuable market share. Because price performance is the ultimate goal of telecom managers, it seems appropriate to consolidate the discussion of the disparate pricing schemes in a single chapter at the end of the book.

Aggregation

Despite the efforts of AT&T, MCI and US Sprint, the marketplace will continue to emphasize price heavily. As communications grows as a strategic resource for user companies and telecommunications costs increase, telecom managers will feel more pressure to reduce expenses.

One market development that started making headlines in 1990 was service aggregation. Service aggregators made it seem like 1984-85 all over again – "We offer greater discounts, sign up with us now." According to a TeleChoice, Inc. (Manchester, CT) survey, there are an estimated 250 groups offering aggregation services, with more being formed monthly. Their impact on the market is incredible. The buying weight these groups carry is significant – the top twenty groups now represent an annual buying power of nearly $1.6 billion. That's more than 3 percent of the $50 billion long distance industry. What's more, the average service aggregator has been aggregating traffic for just over one year.

Checking out an aggregator

User should treat a service aggregator as they would any long distance company. Before signing, telecom managers might ask the following questions to make sure the company is someone to do business with:

1. How long have they been in business? Is this a start-up firm just to aggregate discounted traffic, or have they a solid corporate history in another aspect of telecom?

2. How long have they been aggregating traffic for a business? Older aggregation firms tend to have larger traffic portfolios. This is not always true, however. Some firms have large amounts of traffic because of their other areas of business – for instance, when Equal Access Corp. started aggregating outside customers, they brought with the company its embedded base of 1+ and 0+ customers from its payphone and fax programs.

3. What experience do they have with networking and long distance? Is this just a sales company or do they know something about telecom too? Will they analyze a network BEFORE recommending a solution? Will this cost extra?

4. What carriers do they resell? Some companies just sell AT&T. If a customer wants discounts, then it has to convert to AT&T. Many companies do not want to go with one carrier and would like to be able to stick with several carriers. A truly flexible company will offer a choice. Remember a core reason for this environment – aggregators give AT&T a shot at getting all of a company's traffic.

5. What services from those carriers do they resell? Some aggregators only resell WATS or 800; others will sell virtual networks, travel cards and other local or long distance services. If the aggregator says that it uses a virtual network, user should be sure to ask if it is presently installed (it takes about six months to get a virtual network installed today). Surprisingly few aggregators will help users obtain discounts for private line services, although this could change in the near future because of the huge margins existing in private line term plans. Most will merely help a user place orders for the service through the existing account. In all, users are better off giving all of their traffic to an aggregator who can handle the most of their business.

6. Whose customer is the user? Who does a customer call with a service problem? Is the users still a customer of the carrier or are a customer of the aggregator? This is the most crucial question a user can ask, and the answer can change over time too! Many aggregators will represent to a user that nothing will change about the user's relationship with the carrier. This may or may not be the case. In general, AT&T is shifting its stance over who is the customer, the end user or the aggregator. AT&T used to say that the end user is the customer of AT&T. Now, the end user may be the customer of the aggregator, depending on which service is purchased. An example, if the user is an SDN customer and it shifts to a partition of an aggregator's SDN, that user will lose its AT&T account team and is now a customer of the aggregator – and must use its customer service representatives to deal with its SDN account team.

7. Is the aggregator a sub-aggregator for any other company? This is important. Some aggregators have signed other aggregators to be part of their discount plan. If a user signs with a sub-aggregator for AT&T SDN, that user might have to go through two levels of customer service before a problem is presented to the carrier. Similar problems exist for billing. If dealing with a sub-aggregator, find out exactly what the relationship is and to whom.

8. Who will bill? Are users going to be billed by the aggregator or by the carrier? Again, this depends on the individual service. Some aggregators give the user the choice – standard bills from the carrier or customized bills from the aggregator. The carriers appear to be pushing aggregator billing as a means to limit the number of aggregators in the market. If this is the case, what experience does the aggregator have in billing? This gets back to the question above concerning the outside telecom experience of the company.

9. Who is responsible for the bill? While the user may pay your monthly bill, it is likely that the user not the entity that has overall financial accountability for the account. Typically, this is the aggregator. However, there are instances where the end user could be financially responsible for non-payment by other members. An example is some forms of co-ops.

10. What value-added services does the aggregator offer? Network design and optimization is typically offered, but the level and depth of these analyses vary substantially. Users should find out exactly how the company intends to analyze the bill. Users should not settle for an answer such as "We use state of the art computer technology," and should find out the process and make an assessment of their true capabilities. Also, users should beware of bias. When the firm is selling just one carrier, how much optimization can occur? Look out for solutions that fit perfectly into what the aggregator is selling.

11. What are the installation, monthly and annual costs? Each of the aggregators charge different amounts for their services. Some charge a flat monthly fee, either by account or per location; some charge a percentage of "improved" savings (i.e., a portion of the additional savings resulting from participating in the program). Still another method is charging a flat fee to join and then a variable percentage of the end user's savings based on volume of usage. If percentages are involved, user should be sure to ask how they verify these improved discount levels.

12. Does the company have extra profit sharing arrangements? A few of the aggregators offer unique profit sharing arrangements with their customers in addition to the improved monthly savings. These can range from a percentage of overall annual sales to a fixed sum based on customer size. For instance, Consortium 2000 of California offers their customers an annual "dividend" in addition to monthly percentage-based savings. These dividends are based on total amount of traffic during the previous 12 months.

13. How large a staff does the company have? Is this a one-man outfit or a sizeable company? If this firm is responsible for all billing and customer service for your firm, users should be sure that it has the wherewithal to perform the tasks. If the user has to switch from an AT&T account team to the aggregator's team, what differences will the user firm see in response times? Are there 24-hour, 7-day-a-week help lines? How many people are on the staff? What is the experience level of the personnel? Is there an 800-number for service problems?

14. Is long distance the only thing they aggregate? Some companies aggregate more than long distance. Centex Telemanagement for instance buys Centrex lines in bulk and passes the savings onto customers, adding in telemanagement expertise int he process. Garden State Long Distance sells cellular service. Some companies are general buying clubs, aggregating non-telecommunications items from fax paper to soap.

15. Is the company an authorized aggregator? There are many aggregators in the marketplace that are misrepresenting their relationships with the long distance carrier. In surveying the aggregators, TeleChoice found one company that claimed to be officially reselling US Sprint services, when in fact a cease and desist order had been placed against the firm from aggregating with Sprint services. For a while, AT&T authorized aggregators through marketing agreements. AT&T signed joint marketing agreements with aggregators based on credit worthiness, reputation and integrity, established business plans, marketing methods, size of existing sales force and the anticipated target market of the aggregator. AT&T has stopped this now. MCI and US Sprint do not condone aggregation, but have no problem selling their services to "legitimate" resellers or through agents of the company. Should a company represent itself as tied to a particular carrier, users should verify it with the carrier through the local sales office.

16. Are the bills correct? Certainly one of the often mentioned criticisms of the AT&T aggregator program is billing. According to industry participants, AT&T's ability to accurately reflect discounts on bills has been lacking. If a user has joined or is planning to join any carrier's program, it should be sure to look closely at each month's bill. Users should work with the aggregator to verify all line items and usage costs. It does no good to sign up for a discount plan and then not get the discounts.

17. What about intrastate traffic? Can the aggregator provide intrastate savings? Under some plans, intrastate traffic cannot be aggregated. Also, user should ask about intra-LATA traffic. One aggregator, Consortium 2000, offers this example: regional resellers has been offering intra-California of 12 cents per minute. Consortium 200 cuts this to 9 cents per minute in California, plus additional revenues that come from annual profit sharing plus network optimization. In practice, these can combine to bring rates as low as 8 cents a minute.

18. What are you other options? Users should not assume that a service aggregator is the best option for your firm, because it may not be. Is the user a member of a national association that may also have similar discount plans? Several banking associations, for instance, have member discount plans that can often offer the same discount rewards as an aggregator. User groups, affinity groups, alumni associations – all have been active in seeking long distance discounts for members.

Discounts for the average company due to aggregation can sometimes reach as high as 20 percent. Even some carriers are getting into the act – Cable & Wireless aggregates customer 800 traffic under an AT&T 800 term plan. Aggregating traffic is not a new idea. It is the basis upon which most long distance carriers and resellers got their start – buying bulk and selling piecemeal. Profits come from the "arbitrage" or difference in pricing. But applying the term "service aggregators" adds a new level of confusion to purchasing long distance service.

The Emergence of Aggregators

Aggregators started appearing in response to the introduction in 1988 of AT&T's Multi-Location Calling Plan (MLCP) for outbound WATS service and Revenue Volume Pricing Plan (RVPP) for inbound 800 calling. AT&T originally intended these tariffs to allow customers at the low end of the market to combine their usage to gain discounts they would not be able to achieve individually, in cooperation with Federal Communications Commission rulings on nondiscrimination of service pricing due to customer size.

The presence of aggregators is due in part to anomalies of the marketplace. When AT&T originally filed these discount plans at the FCC, the carrier did not foresee that they would be used by aggregators. Innovative telecommunications entrepreneurs have taken advantage of the multisite offerings to form aggregation groups.

Each of the carriers gives users higher discounts for greater levels of monthly usage. At the same time, the tariffs allow companies with multiple locations to group their traffic together to obtain these higher discounts. This is what makes service aggregation not only possible but unlimited. Because the carriers are bound by the FCC to make their products available to all businesses in a nondiscriminatory manner, it is difficult to allow a company to aggregate traffic from its seven satellite sites under one plan and not to allow a service aggregator to group seven unrelated sites under that same plan.

Aggregators are advantageous to the carriers. They allow a carrier to protect a base of price-sensitive customers by offering a more competitive service option than that which normally might be available to an end user of a particular size. Aggregators also enable carriers to go after all of a customer's traffic by bringing all sites and traffic under one pricing plan. This can help AT&T in its fight to win back lost accounts.

Aggregators also have a degree of expanded market coverage that is unrivaled by existing sales forces. Aggregators often make use of existing contacts and client bases to market services, sometimes reaching outlying locations where carrier representatives and marketing programs rarely go. Long distance carriers have relied for years on outside third party marketing firms to sell their products nationwide. For instance, MCI forged an early relationship with AMWAY, famous for its door-to-door sales force selling everything from brooms to shampoo. Under third party marketing, the carrier provides the service and sometimes the billing, the outside marketer does the selling, using cost margins as a source of revenue.

The Limits of Aggregation

Most aggregators offer exclusively AT&T services. Because AT&T has more than 75 percent of the total network services market, it makes more sense that an aggregator would choose to aggregate with AT&T's services. While MCI and US Sprint have been working in the legal, sales, and regulatory realms to inhibit service aggregation, AT&T has been aiding the resale of AT&T services. For a time, AT&T was signing marketing agreements with some of the aggregation firms, claiming other carriers were ignoring certain types of customers in the market – customers that clearly have a need for these sorts of services. AT&T has at least six different multilocation, and in some cases multiservice, discount plans that service aggregators can use to obtain higher discounts for their traffic.

However, problems have quickly arisen as aggregators target AT&T's larger accounts, creating "channel conflicts" in the sales organization. After all, the fastest way to get to higher discount levels is by signing large volume accounts. Predictably, AT&T salespersons have been upset. Aggregators sometimes outbid carrier sales personnel using that carrier's own products but with better pricing and more flexible contract terms.

AT&T has made it clear to its outside marketers that there are certain areas that are off-limits to aggregation. Large accounts are one such area. AT&T has gone so far as to make some of its products less attractive to high-end aggregation through various rate changes. One set of 1989 adjustments to the 800 revenue volume pricing plan which substantially reduced the margins in high-volume aggregation of toll-free services was explicitly intended to discourage customers with 800 services tariffed charges greater than $50,000 per month from seeking discounts through an aggregator. Similar changes to SDN have made that product unattractive to residential aggregation.

AT&T would like aggregators to target customers in the range of $500-$5000 usage per month. Indeed, the survey conducted by TeleChoice found that the average customer of an aggregator spends $3,650 on long distance monthly.

MCI and Sprint Shun Aggregation

MCI and US Sprint have taken a remarkably different tack in approaching aggregation. Neither have problems dealing with resellers or rebillers, but both have taken steps to ban pure aggregation of their services. MCI and Sprint already market their services as lower in cost than those of AT&T; both likely have lower operating cost margins than AT&T, making it harder to create the same arbitrage situations as AT&T. Aggregators are naturally attracted to AT&T's higher margins.

Surprisingly, it is in the mundane language which defines each tariff that this battle against aggregation is being fought. MCI and US Sprint have adjusted their definitions of "customer" to rule out nonrelated businesses. Look at some tariff language for one of MCI's services: A customer is defined as "a trade association representing business entities or individuals within an industry, professional or business classification, or a commercial organization with affiliated franchises, independent agents, independent distributors, or other multiple commercial representatives, or a buying group not organized for the purpose of qualifying for the discounts provided for herein."

Even with these guidelines, it is tough to define away service aggregators. Both MCI and Sprint have complained to the FCC about AT&T's attitude on aggregation, complaints AT&T has dismissed as unwarranted. Ultimately, the FCC will decide on whether or not AT&T can legitimately promote aggregation through its tariff structures. At a time when the FCC is considering the release of AT&T from regulation, a ruling barring such practices is not likely.

One of the most controversial issues before the commission is MCI's claim that AT&T is targeting MCI and Sprint accounts through these aggregator programs. MCI's filing gives examples of contractual requirements dictating specified levels of traffic won back to AT&T in order for an aggregator to remain a part of particular third-party marketing programs. AT&T has said that it certainly encourages such win-backs, but has not specified exact levels for each aggregator.

The Regulatory Status of Aggregators

Service aggregators are walking a thin line between classification as resellers and mere marketing agents for the carriers. Several states are reconsidering just how these entities should be regulated. Regulated resellers and rebillers are pressuring state PUCs to regulate aggregators as well. In the absence of equal regulation, these long distance companies are feeling the squeeze to convert to service aggregators themselves – something the supplier carriers definitely want to avoid.

Some service aggregators are being closely watched by PUCs for other business-related regulations. For instance, the state attorney general's office in West Virginia recently sued one aggregator, claiming the company had tried to start a "pyramid" marketing scam using service aggregation. Indeed, company brochures explained how salespersons could get a new Mercedes Benz and free trips for two to anywhere in the world by signing up additional sales personnel.

In all, users are best off to select an aggregator that has a history of experience in the telecommunications industry and who is open to a range of carrier and service solutions. These companies are in the business for the long run and will not abandon customers when the tariffs change.

The Future of Aggregation

For now, service aggregation is the most cost effective option for most AT&T users. By signing with the right firm, an end user can boost discounts and possibly pick up value-added services as well.

However, the whole market is changing rapidly. All of the carriers are changing their policies and plans "on the run" – trying to decide whether to allow or continue, or actively to encourage aggregation of services. Competitors are watching AT&T closely to determine the carrier's future direction in this market. AT&T appears to be reconsidering the long-term impact of its moves. In April of 1990, the carrier began publicly denouncing aggregation and made the first moves to halt the phenomenon wherever it could.

The concept of buying bulk and selling piecemeal is not going to disappear. None of the long distance carriers are opposed to "legitimate" reseller and rebiller operations, nor are they opposed to sales agency agreements for their products. MCI, Sprint and a host of other long distance carriers are opposed to service aggregation.

In the future, users will see adjustments to the market. AT&T is adapting to the impact of service aggregation on its product and pricing plans. Channel conflicts cannot continue to occur and at some point, AT&T will impose constraints on third party aggregation. At that time, AT&T may join the other carriers in discouraging service aggregation in certain market segments. Among the options AT&T has are:

1. **Tariff modifications.** Changing tariffs further to discourage aggregation of certain segments of the marketplace – areas where AT&T is confident of its market share.

2. Account team limitations. Limiting or removing direct customer support by forcing customers to deal with an aggregator first.

3. Aggregator migration. Migrating aggregators to a full alternate channel role in which they are AT&T trained and compensated for sales to AT&T in select target markets. This resembles more of an agency approach to the problem.

For the time being, the absence of tariff language prohibiting nonaffiliated customers from taking advantage of AT&T multilocation discount plans should not keep a company from joining an AT&T buying group. It will be harder to find a Sprint or MCI buying group that is not a reseller or rebiller.

Right now, AT&T at least wants to keep the aggregation issue alive. It has been intent on stopping erosion of its customer base and sees aggregation as one means to that end. Should AT&T further restrict the tariffs or marketing agreement programs, the existing aggregators will have to decide whether to change marketing plans, make the jump to a full resale or rebill status, or to politely bow out and await the next telecommunications opportunity.

Meanwhile, the market is open for new aggregators – any entrepreneur who has decent sales ability and is willing to take risk can become one easily. However, competition among aggregators will reduce margins and force many to launch value-added services. Look for consolidation as aggregators get aggregated themselves.

The Move to Switched Services

Another major pricing trend in the market involves the move to switched services. The long distance carriers want to migrate users from private line networks to bulk voice services – such as Megacom, Prism I, Ultra WATS, SDN, Vnet and VPN – where the long distance carriers can enhance revenue streams while more efficiently utilizing networks.

When a carrier sells a user a private line on a monthly basis, the carrier releases all claim to using that line – even if it is not being used by the customer at all times. Few users are able to pack traffic on a private facility 24 hours per day. Even so, the carrier cannot send any other traffic over that facility. For this reason, the carriers want to promote switched rather than private facilities. When a circuit is not being used by one customer, the carrier can use it to carry another customer's traffic. This increases the efficiency of the network and raises profits.

The carriers are using virtual network services, with their complete network management capabilities, including routing and maintenance management, to tempt users away from private lines. The carriers want to convince private network users that it is both uneconomical and impractical to "run your own telephone company."

The carriers are taking advantage of the fact that many companies are tired of managing their own large corporate networks. These companies want simplicity from their telecommunications systems. Managers can eliminate the hassles of implementing, maintaining, and upgrading a private network by handing the whole chore over to a carrier. Also, companies do not have to buy expensive new technologies as they become available because the carriers will be installing them in their own networks – and launching them as features or products available to users.

Once enough customers have migrated to switched services, it will be difficult to return to private line networks. Should the carriers decide to raise switched prices, it would be expensive to recreate private line networks. The cost of leaving the virtual network to return to a private alternative would be much higher than bearing the greater expense of staying with the virtual network.

AT&T's Strategies

AT&T has shown with its cascade of tariff filings at the FCC that it intends to be price competitive to prevent losing any more market share. Where AT&T cannot sign long-term contracts, it is trying to mute the differences between its products and competing ones. AT&T wants the momentum in its client base and its reputation, rather than simply price and features, to become a greater determinant in the customer's selection of a carrier.

As AT&T's regulatory environment becomes less stringent, the carrier will increasingly use price as a means to get and keep customers. The latest round of FCC approvals for pricing plans and price cap regulation shows that the FCC is letting AT&T have its way with pricing. With more than 30 percent of its overall market share eroded, the carrier is claiming that it is no longer the dominant carrier in the marketplace, and internal government reports are supporting this contention. With AT&T on the loose, competitors are preparing for an onslaught of price battles. AT&T has five major pricing weapons.

The first are special tariffs. AT&T has introduced several new tariffs – Tariff 12, Tariff 15, and Tariff 16 – with more promised in the near future. Tariff 12 applies to custom-designed networks for users, allowing AT&T to offer special pricing and services. AT&T claims to have more than 100 customers waiting for special tariff deals.

A second pricing method deals with volume pricing plans. For all of its products, AT&T wants to offer a complete, cost-effective service structure for all companies, regardless of their needs – the "total solution" so to speak. The company has introduced a score of interconnected plans throughout their pricing structure, in order to present a unified telecommunications package. However, AT&T has been careful to implement these wide-ranging discount plans in a stepwise fashion, making changes to each of its product lines individually to reduce the expected regulatory resistance from the FCC and competing carriers. Had AT&T tried to restructure all of its services at once, the FCC most likely would have struck down the plan.

A third pricing method is the manipulation of pricing elements to direct customers further down the AT&T product line. For instance, AT&T lowered the installation fees for direct connection to SDN to have as many customers on dedicated facilities as possible, largely because ISDN requires dedicated connections and AT&T touts its is pushing ISDN as a distinct advantage over the competition. AT&T's network is nearly 90 percent digital and AT&T will be able to provide ISDN capabilities nearly network-wide by the end of 1990. AT&T is now starting to provide geographically dispersed customers ISDN features such as station identification and 64 kbps data transmission. To prepare its customer base for these products, AT&T has been putting them on dedicated facilities.

AT&T also filed select promotions that waive for a limited time most installation fees for various long distance services, including Custom SDN and many optional features. The waivers cut to the heart of Sprint and MCI marketing plans which thrived on AT&T's high service installation fees.

The fourth method is term plans. Most of AT&T's major products now have one-to-five-year plans associated with them. AT&T has tied lower installation fees and higher discounts with its term plans to take advantage of that portion of the marketplace that cannot afford installation fees but want AT&T service.

The fifth and newest method is promotions. AT&T has been filing special promotions for various services, offering waivers of installation fees, feature charges, minimum requirements, etc. These have had a profound effect on the market. AT&T's autumn, 1989 promotion waiving installation fees on SDN generated so many responses that a six-month backlog on installation developed. Although Sprint and MCI could dominate the low end of the market in the past because low install and feature fees, AT&T's promotions have taken away that edge – exposing MCI and Sprint to the full competitive force of AT&T's products. In many cases, these promotions have served as the deciding factor in the selection of services.

Migratory Product Line

AT&T uses these pricing methods selectively to control the direction of demand for its products. No one product can be seen in isolation. Taken together, all of AT&T's pricing plans create a migration path for their long distance customers. Consider a multisite company which initially signs with AT&T for a range of Pro WATS, Megacom, and 800 services. Under the Multi-Location Calling Program, the user can attain some savings among its various locations. As the company grows and maybe acquires other companies, the amount of on-net traffic grows and the user migrates up to a Basic SDN.

AT&T has already built this "bridge" into the MLCP tariff – a user migrating to an SDN can terminate the MLCP without penalty. As the customer becomes familiar with SDN and as the needs of the firm grow, the company moves up to a Custom SDN.

Beyond Custom SDN lies AT&T's customized offerings as provided by Tariff 12. When combined, these services complete the entire chain of switched access discount plans. Naturally, if needed, a special tariff can be filed to accommodate any problems.

One cannot understate the effect that regulation has had on the development of AT&T's services, particularly SDN. AT&T designed SDN to mimic existing plans in order to create as little clamor at the FCC as possible. A lot of features and options that AT&T could provide are not available to the general public largely because the regulatory people have said, "If you don't want to make it available to everyone, you can't do it." Put in this position, AT&T has introduced some service features rather slowly. However, as the FCC becomes more lenient towards AT&T and allows the company a more competitive response to market conditions, AT&T will become more responsive to customers.

AT&T Network Redevelopment

Recently, AT&T has started to feel the effect of the competing carriers' network strength. The carrier has lagged behind in providing some of the more advanced network offerings because of old analog equipment that was still in place throughout its network. However, AT&T has invested more than $8.7 billion between 1988 and 1990 to improve its core network. The carrier has accelerated the digitization program of its domestic and international telephone networks, largely in response to the high-visibility network reliability and quality issues stressed by competitors. AT&T

plans to continue this investment at the rate of $2.5-$3 million per year over the next several years.

AT&T is increasing the level of intelligence in its network to allow for "self-healing" – whenever a major disruption occurs, rerouting can take place almost instantaneously. This feature is will allow for greater switched data transmission through the network.

To help increase sales and improve responsiveness, AT&T redeployed more than 5,000 managers into the field to sell AT&T network services in 1988, 1989 and 1990. The carrier is attempting to shed its image as inflexible and unresponsive to customer needs. The managers are providing customers with strong technical support in handling everyday network problems.

In all, with more people in the field, a better network to sell, and pricing plans that can be molded to win a particular bid, AT&T looks strong going into the next decade. MCI and Sprint will be hard-pressed to fulfill their promises of beating AT&T's features, reliability, and core network intelligence for virtual network services.

US Sprint's Strategies

US Sprint and MCI are both targeting the same market as AT&T. Sprint made its initial mark in the low- to medium-end of the market, but recently has signed some of AT&T's former high-end customers to its ranks.

Sprint stresses its networking experience and fiber facilities in its sales promotions. Not one Sprint press release has failed to mention the company's "100 percent all-digital fiber optic network." The carrier has placed most of its sales hope for the upcoming years in this network. Sprint completed its national network development program in 1988. The carrier has three diverse coast-to-coast routes for redundancy in its network, and has added sophisticated monitoring equipment which helped to win part of the federal government's FTS-2000 contract. It was the first carrier to install SS7 throughout its network. Sprint proudfully touts this victory in its promotional campaigns, with good reason. The carrier stands to gain considerably from such a large addition to its base of embedded customers.

Sprint sees itself as having leapfrogged the competition with regard to network capabilities. AT&T and MCI are to some degree "playing catch up" to match Sprint's all-digital facilities. Sprint originally thought its "window" on network supremacy would be two and a half years, but accelerated AT&T and MCI programs have cut this to one to one and a half years. Sprint will use this time to try to catch up itself in terms of feature availability and to offer more enhanced services, such as subrate T-1 services and other digital services that will build on the strengths of its network. As of the middle of 1990, Sprint is moving quickly toward matching the MCI and AT&T products feature-for-feature, aided in a large part by its all digital, all fiber network architecture. The network allowed Sprint, in July 1989, to introduce its VPN 56 switched data capability at voice rates, forcing MCI and AT&T to respond in kind.

Sprint is expanding its core network capabilities in an effort to match those of AT&T and MCI. At the end of 1989, the carrier added a range of extended-digit screening capabilities that allow for enhanced processing of 800 and virtual network calls. This one function is critical to a series of features for VPN.

One problem Sprint has struggled with for years was billing. When GTE Sprint and US Telecom merged, the resulting integration of their combined billing system nearly crippled Sprint. In early 1990, Sprint introduced its new billing program that greatly enhances its ability to offer special call detail reports and multilocation special deals. Of particular importance is the integration of the data network services under this new billing system – that will allow across-the-board pricing plans for voice, data, image and facsimile. The big issue will be implementation of that new billing system, something that Sprint had problems with in mid-1990.

Sprint stands to reap tremendous benefits from its international relationship with Cable & Wireless. The other carriers will have a hard time matching the resulting international virtual network and other services characterized by integrated intelligent network systems.

Furthermore, Sprint is reorganizing so that it manage its international and data divisions in a more integrated fashion by merging the two groups into the core Sprint organization. The carrier's parent company has even renamed itself – from United Telecommunications to "Sprint" – to show a more united front against the competition. The name change was formally delayed until 1991 due to a marginal financial results in the second quarter, 1990.

Where Sprint may fall short is in the support structures that bolster new features and programs. The carrier's limited ability to program additional billing modules into the invoicing system will largely dictate the rate of implementation of new features. While much of the necessary programming for the billing of future products is already in place, some of these modules have yet to be designed. Sprint has no intention of losing the war with AT&T merely because of delayed feature deployment. The carrier will attack AT&T head-on in a battle of price. Sprint has long used innovative pricing to win customers, and will continue to do so.

MCI's Strategies

MCI vastly expanded its intelligent network capabilities, its data transmission services, and its network management capabilities in 1989 and 1990. MCI wants to differentiate itself from AT&T and Sprint through its distinct network management offerings – giving users the ability to access and control their own networks, private, virtual or hybrid.

MCI has started implementation of its advanced integrated network management system (INMS). The INMS package will anchor MCI's intelligent network. It will provide the user with the ability to monitor, analyze, and control any part of the network from a single terminal. For instance, a user can monitor the status of Terrestrial Digital Service (TDS) 1.5 lines, reassign Vnet circuits, order more MCI 800 circuits, and get on-line call detail for the Prism services – all from one on-site terminal. Unlike AT&T, MCI will give the customer the capability to manage its network.

MCI is emphasizing the integration of voice and data. Sprint is also capable doing this right now with its network, but MCI is taking its digital functionality across product lines – DDS, Vnet, Prism, 800 – and creating one unified product line. MCI calls this its "Digital Gateway."

MCI has been quite flexible in meeting customers' requirements. MCI must now package the specialized applications it has developed for a few customers and offer them to all customers. The problem with officially launching these services nationwide is that many of the same people who would package and test service feature enhancements are working on the high-priority INMS product. MCI is sometimes too lean an organization.

MCI is arguing that the battle of physical networks is over and the battle of intelligence and software has begun. In reality, of the three carriers, MCI's network is the weakest; however, this is balanced by the fact that its network intelligence is perhaps the strongest. MCI is continuing to improve the core network upon which this intelligence is based. The carrier has finished SS7 implementation. It is continuing to replace analog transmission facilities with digital equipment. The acquisition of Telecom*USA will greatly improve MCI's fiber facilities. The company still maintains that microwave transmission is desirable in certain situations and will therefore keep its digital microwave network intact.

MCI is working to raise its network to the next level of sophistication – ISDN. The company is promoting not just the unified standards of ISDN, but also the intelligence capabilities that ISDN will provide – specifically, customer control of the network. Bandwidth on demand, closed user group signaling, central rerouting, network management and alarming, operations performance and reconfiguration – all of these are features that ISDN can provide through the enhanced intelligence of the network.

MCI is committed to maintaining its traditional pricing position relative to AT&T. The carrier has a stated objective of prices at least 10 percent below AT&T's. Functionally superior at a lower cost is the concept behind their marketing campaign.

The Future of Pricing

In all, AT&T, MCI, and US Sprint are likely to continue to dominate the long distance market. While other carriers may launch regional or even national services, it is likely that these will be limited to offering enhanced WATS and private line services more than anything else.

The top three carriers have low-cost facilities based networks which enable them to compete more effectively on cost in the long run. At some point, enough competition will have fallen out of the market that pricing will stabilize. This time may come sooner than most people think.

Chapter 17

Conclusion

The long distance marketplace could serve as a textbook example of the different stages of an industry lifecycle. The modern competitive long distance market began in 1984 with divestiture and the subsequent court decisions that enabled the IXCs to compete head-to-head with AT&T on somewhat equal footing.

The early years were characterized by intense price competition, innumerable providers, and a great deal of customer confusion. Long distance carriers routinely offered rates 40 to 50 percent below those of AT&T. Like the gold rush, hundreds of telecommunications entrepreneurs flooded the market, leasing AT&T circuits and starting long distance companies overnight.

As the first two years passed, the industry expanded and technological innovations were introduced – allowing market leaders with deep capital financing to fund massive expansion programs. Gradually, the distance between the market leaders and the rest of the pack widened. Some of the long distance entrepreneurs dropped out of the race or were swallowed up by the larger carriers.

As the industry continued to grow, customers became more educated and applications expanded. Technology continued to develop at a breakneck pace. Long distance became more than just 800 or WATS service, but a range of other long distance services. Long distance usage soared, prices continued to plummet, and more long distance providers dropped by the wayside.

Recently, the market has started to reach its first peak. The recent announcement by MCI that it would acquire Telecom*USA demonstrates that the leaders are continuing to distance themselves from the remaining providers. Consolidation is still taking place at the high end of the market and the remaining players continue to fight for hard-to-define niche markets.

Price competition has driven long distance rates to near their underlying costs, favoring the low cost providers. However, as the number of competitors has decreased, pressure on price competition has partially let up. Some prices have actually begun to rise.

The market appears to be settling down somewhat. There will be more consolidation at the top as the remaining struggling players find partners. Meanwhile, operating efficiencies from the earlier capital expansion programs are now giving the market leaders the edge in the marketplace. For the first time, some of AT&T's competitors are regularly reporting quarterly profits.

The New Long Distance Environment

The carriers are moving from mere providers of long distance service to strategic partners with business. Telecommunications is realizing its place in the global economic and political infrastructure. International communications will come into focus quickly. The upheaval in the Eastern European countries and the movement towards a unified European market in 1992 is creating an expansive economic environment that will be attractive to a vast number of international corporations. As the boundaries come down, a whole new marketplace is opened to providers in the domestic U.S. local and long distance market. In this new age of growth, the number of players will be limited from the start, by international operating agreements, trade regulations, and politics. The catch phrase will be "growth through global partnerships."

Finding the User

Where will the individual user in the U.S. fit into this future marketplace? The market is going to change much to the user's benefit. The carriers are moving more towards a service industry, in which customer service, technical support, and enhanced services will play an increasingly important role – more so than basic service pricing. Again, the idea of a strategic partnership between the telecommunications professional and the service provider comes to mind. More and more, customers will rely on their carrier to provide innovative ways to increase revenues and decrease costs – not just in telecommunications but throughout the corporation.

Such items are already being seen in the market. Videoteleconferencing increases communication and cuts travel costs. Toll-free automatic number identification, in combination with demographic survey information, allows market penetration studies. Carrier 900 services provide enhanced revenues and interactive customer response. Newer technology-driven services are on the way – ISDN, bandwidth on demand, etc.

For the end user, the emphasis on new applications, heightened technical support and expanded customer service will mean more return for each telecommunications dollar spent. Although pricing will continue to be a major issue for long distance users and providers alike, it is likely that increasing technological advancements will bring to fruition the idea of postalized telecommunications units – single incremental rates for all types of communications. For instance, it is not fantasy to think of a day soon when all inbound or outbound calls to anywhere in the U.S. will cost just 1 cent per six second increment – with no mileage based cost elements, time of day limitations, or data versus voice considerations. The market is heading this direction rapidly, which is why the carriers are stressing nonprice based factors in the buying decision.

As competition moves the industry towards a national oligarchy, pricing will likely stabilize, at least until the BOCs are released by the courts to provide interexchange long distance services. Nevertheless, there is not enough room in the market for seven new long distance carriers – BOC and IXC partnerships will be fashionable, such as NYNEX and Sprint or Pac Tel and MCI.

In the end, however, users should look to get more from their long distance carrier for the same amount of money, not greater discounts for the same service. Applications and vertical marketing will drive the industry over the next five years and decreased competition will stabilize prices.

It will be exciting to see the proliferation of integrated telecommunications services at each user's fingertips. The future promises of applications-rich home terminals and full-featured portable communications are closer than ever. Today's notion of "long distance" will become obsolete, as communications with the other side of the world becomes as simple and inexpensive as calling down the street.

Glossary

The following definitions of telecommunications terms, acronyms, and abbreviations apply for the meanings used in this book. Some of the terms may have additional or more general meanings outside of this context.

abbreviated dialing – This feature of PBXs, BOC services, or various long distance services enables a user to dial a two- to four-digit code that will be translated into a full ten-digit or international telephone number passable by the public switched telephone network. Typically, the telephone numbers that are represented by the abbreviated numbers may be programmed and adjusted at will by the user. Also known as speed dialing or speed calling.

A block/B block – These are the two sets of channels assigned by the U.S. government to the cellular telephone companies awarded operating licenses in each metropolitan area.

access – The ability to enter a network.

access charge – A fee levied by the local exchange company for the use of its local exchange network for the origination and termination of calls. (See also switched access.)

access line – The circuit that connects a user location to a switching center.

Access Line Grouping Call Screening – A feature under AT&T's SDN virtual network service, a customer can specify calling privileges for on-net stations on an access line group basis. These may contain up to 200 different definitions of area code, exchange, or station groups.

access tandem – A telephone company switching system that allows an interexchange carrier to make originating and terminating connections at multiple end offices within a single LATA.

account codes – In various long distance services, these typically are two- or three-digit codes that are dialed after the telephone number, which allow allocation of call costs to specific billing entities.

Accunet – AT&T's trade name for a group of digital data transmission services that use terrestrial transmission facilities.

Accunet Spectrum of Digital Services (ASDS) – AT&T's fractional T-1 private line offering that allows users to buy digital capacity at a portion of the cost of a full T-1 circuit.

Accunet Switched 56 Service – AT&T's trade name for a digital full-duplex data transmission service, operating at either 56 kbps or 64 kbps, that offers substantial digital capacity over a terrestrial facility network on a dial-up basis.

Accunet T45 Service – AT&T's trade name for an ultra-high capacity backbone transmission service. It is a digital service employing terrestrial fiber optic lines to transmit voice, data, and video at 44.6 Mbps. Available in major business centers where 44.6 Mbps fiber optic local access facilities are available.

Accunet T1.5 Service – AT&T's trade name for a high capacity digital transmission service that uses terrestrial 1.544 Mbps digital private lines for point-to-point connections for simultaneous transmission of voice and data. The lines can run between two subscriber points, between the subscriber and an AT&T serving office, or between two AT&T serving offices.

Accunet T1.5 Bandwidth Management Service (BMS) – Allows customers to reconfigure and restore high capacity digital networks in less than a minute, as well as perform customer-initiated testing and oversight network performance and alarms.

Additional International Discount Plan – An AT&T discount plan that gives SDN users an extra discount on customer-dialed calls to more than 40 countries, for all remote and local exchange access SDN users.

ADPCM (Adaptive Differential Pulse Code Modulation) – An often used alternative to PCM (pulse code modulation) that is used to convert analog signals to digital. The biggest advantage of ADCPM is that it allows two 32 kbps channels to occupy one, industry-standard 64 kbps channel. The amplitude of the analog signal is sampled 8,000 times per second and the difference in value between successive measurements is expressed as a four-bit binary number. Transmission of 8,000 four-bit binary numbers each second corresponds to a 32 kbps rate.

Advanced 800 service – 1) AT&T's trade name for enhanced basic 800 service. There are eight options which are intended to tailor 800 service for the needs of each individual customer. 2) Generic name for advanced 800 service features that were required by Judge Greene in ordering the BOCs to implement SS7 throughout their regions.

airtime – The amount of billable time accumulated by a user, usually measured in minutes.

alpha test – A pilot testing program for unreleased services and products. The test is conducted by the manufacturer to assess performance of a new product or system. The test occurs in laboratory simulations of conditions that the product or system will encounter in real use.

alternate routing – Sending a call over a different route when the first choice route is busy.

Ameritech – One of the seven regional holding companies (RHCs). At divestiture, Ameritech was given control of Indiana Bell, Illinois Bell, Michigan Bell, Ohio Bell and Wisconsin Bell.

analog – A transmission method that is based on a continuous wave electrical signal as opposed to the 0/1 structure of digital transmission. Analog transmission is considered less desirable than digital transmission. (See also digital.)

analog channel – A communication circuit that can send and receive an analog signal. Typically analog channels are used in the transmission of voice and low-speed data communications.

analog switch – A piece of equipment that is used to set up communications paths for the transmission of analog signals. (See also switch.)

ANI – See automatic number identification.

answer supervision – A signal sent through the network that denotes the beginning and end of a phone call. A signal is sent from the terminating switch to the originating switch when the called party answers the phone. Likewise, a signal is sent when either party hangs up. This signaling in critical for both the local exchange company and the long distance carrier in being able to determine the actual billing time for a call.

architecture – See network architecture.

area code – The first three digits of any standard ten-digit public switched telephone number in the North American Numbering Plan. The three digits tell the network to terminate all long distance calls with that prefix to a certain geographic area. The first digit is a number between 2-9, second is always 0 or 1, and the third digit is any number 0-9. Area codes do not always correspond to LATA boundaries, and can overlap LATAs, or encompass more than one LATA. Area codes never cross state borders. (See also LATA, NPA, numbering plan.)

asynchronous – A data communications transmission method in which each character of information is sent one at a time, separated by start and stop elements at the beginning and end of the character. Asynchronous transmission is characterized by irregular time relationships. Also called Start Stop Transmission. (See also synchronous.)

AT&T (American Telephone & Telegraph) – Prior to 1984, the owner and operator of the nationwide Bell system. After divestiture, AT&T was largely limited to long distance services, although the company diversified into other businesses. AT&T is the largest of the long distance carriers with approximately 70 percent of the market. Among AT&T's other business interests: telephone network equipment, customer premises equipment, and computer equipment and systems. (See also Bell System, divestiture.)

AT&T Enhanced FAX – AT&T's facsimile service that allows users to better manage facsimile traffic.

authorization code – In various long distance services, multidigit codes that cause the network to perform extra processing of a call. They are a series of digits that allow a user to gain access to certain facilities, services or features of a network. Authorization codes are usually used with 800-access to virtual network systems. Authorization codes provide increased security on calls and can also be used to direct billing to specific cost centers.

automatic number identification (ANI) – The identification of a calling station's telephone number provided by network signaling at the time of a call. Typically used by carrier for the routing and billing of a call.

B channel – A "Bearer" channel that consists of 64 kbps digital capacity standardized for use by ISDN. B channels may be circuit-switched or packet-switched. B channels are used for digital transmission of voice, data, facsimile, and freeze-frame video. The 144 kbps ISDN basic rate interface (BRI) standard specifies two bearer channels; the 1.544 Mbps primary rate interface (PRI) standard specifies 23 B channels. (See also basic rate interface, D channel, integrated services digital network, primary rate interface.)

backbone – The primary foundation of a communications network that connects major nodes.

bad debt – Lost profits due to uncollectible charges. As regards 900 services, a carrier typically will charge an extra percentage or per call fee to cover bad debt from 900 programming.

band – 1) Used as geographical boundaries defining billing rates for traditional WATS services. 2) In rates, typically a set of upper and lower limits for a particular measurement, such as mileage, hours, or usage amount.

banded WATS – A WATS service in which pricing is based on which geographical band circuit a call was placed. This is opposed to virtual banded WATS. (See also virtual WATS.)

bandwidth – The range of frequencies that can be sent without distortion over a transmission medium, equal to the difference between the highest and lowest frequency in the band. Circuit capacity is measured in bandwidth – the more bandwidth available, the more traffic that medium is able to carry. Bandwidth is measured in hertz.

Bandwidth Management Service – See AT&T T1.5 Bandwidth Management Service.

barge-in – Also known as "listen in," this form of 900 service allows callers to call an on-going message or program. A popular usage is GAB lines in which multiple callers are speak on one line. (See GAB.)

base rate – The standard rate that is charged for a telecommunications service during nondiscounted conditions, from which time-of-day discounts, volume discounts and other rates are calculated.

basic exchange telecommunications radio – See BETR.

basic rate interface (BRI) – The ISDN interface standard for single-line ISDN service, as well as to connect single line telephone or desktop terminal equipment to multiline ISDN business phone systems. The standard provides for 144 kbps of bandwidth – two message-bearing 64 kbps "B" channels for voice, facsimile, freeze-frame video and data, and a 16 kbps "D" channel for network signalling and data. (See also B channel, D channel, integrated services digital network, primary rate interface.)

basic service element (BSE) – A term in Open Network Architecture representing a local exchange access function or interface, or some sort of local service capability that could be used to create new services. (See also basic serving arrangement, enhanced service, ONA.)

basic serving arrangement (BSA) – A term in Open Network Architecture representing the actual transmission facility and central office switching functions sold by a local exchange company. Through the basic serving arrangement, an enhanced service provider can use the unbundled telephone network functions known as basic service elements. (See also basic service element, enhanced service, ONA.)

baud – A unit that corresponds to signalling speed in data communications. The speed in baud is the number of times per second that a signal is altered. (See also bit, bit rate, bps.)

B block – see A block and B block

bearer channel – An ISDN B channel (64 kbps) for transmitting voice, data and images between points. (See also B channel, ISDN.)

Bell Atlantic – One of the seven regional holding companies (RHCs). Bell Atlantic assumed control of the Chesapeake & Potomac (C&P) Telephone Company of Virginia, C&P of West Virginia, C&P of Maryland, C&P of the District of Columbia, Bell of Pennsylvania, New Jersey Bell, and Diamond State Telephone Bell operating companies (BOCs), after the breakup of AT&T.

Bellcore (Bell Communications Research) – An organization that is owned jointly by the seven regional holding companies that conducts research, determines industry-wide standards for the telecommunications industry, acts as a centralized technical resource and provides management services to the RHCs.

Bell operating company – See BOC.

Bell South – One of the seven regional holding companies (RHCs). Bell South assumed control of Southern Bell and South Central Bell, after the break up of AT&T.

Bell System – An term that encompassed the whole of AT&T before the January 1, 1984 breakup. The Bell system consisted of the 24 Bell operating companies (BOCs), which provided local exchange phone service; the AT&T Long Lines division, which provided long distance connections; a manufacturing arm, Western Electric; and a research and development division known as Bell Laboratories. The Bell System was broken up by the AT&T divestiture. (See also AT&T, BOC, divestiture.)

BER – See bit error rate.

beta test – The last phase of testing program for an unreleased product. In a beta test the product is given to selected commercial customers to evaluate how the product or system performs under everyday conditions intended for use. Also known as field trial. (See also alpha test.)

billing – In long distance jargon, this is actual end-user billing for a call. (See collection).

bisynchronous – A character-oriented IBM data communications protocol.

bit – Abbreviation of binary digit. The smallest unit of digital information utilized by electronic or optical information processing, storage or transmission systems. The binary system is based on the numbers 1 and 0. These digits correspond to the off and on states of a digital computer logic element or one of two possible discrete states within computer memories or digital voice-data transmission systems.

bit error rate (BER) – The number of bits unintentionally changed during data transmission with respect to a specific quantity of bits transmitted. Bit error rate is usually expressed as a number referenced to a power of 10.

bit rate – The transmission speed of bits. Expressed usually as bits per second (bps).

BOC (Bell Operating Company) – The common term for one of 22 local exchange companies that were part of the Bell system prior to divestiture. Each of the BOCs is owned and managed by one of the seven regional holding companies (RHCs). Approximately 80 percent of America's local exchange users are served by the Bell operating companies.

bps (bits per second) – The primary measuring unit of speed in a digital transmission system. The number of bits that a digital transmission facility can convey between a sending location and a receiving location in one second. Rates can be expressed in thousands per second (kbps, kilobits per second), millions per second (Mbps, megabits per second), billions per second (Gbps, gigabits per second), and trillions per second (Tbps, terabits per second). Also noted as b/s, or bit/s.

broadcast – As regards facsimile network services, the ability to send the same message to multiple locations on a near simultaneous basis.

BRI – See basic rate interface.

BSA – See basic serving arrangement.

BSE – See basic service element.

busy hour – Occurs when an uninterrupted 60-minute period of maximum network volume is achieved.

bypass – The substitution of any circuit configuration for the switched services or access facilities of the local exchange company. One type of bypass is "service bypass," which occurs when a customer uses dedicated telephone company circuits instead of switched services. Another type is "facilities bypass," which occurs when a customer directly connects his premises with a long distance carrier's point of presence for the purpose of not using the carrier access services of the local exchange company.

byte – The smallest unit of information that a computer system can locate within its data storage or memory. A byte typically consists of eight bits and represents an amount of information roughly equivalent to a single printed or typewritten character. In special-purpose data storage or memories, a byte may consist of more or fewer than eight bits.

call forwarding – An optional feature that allows the user to program a phone to ring at an alternate location; call forwarding may be set up to be in effect at all times or just when a particular phone is busy or does not answer.

call hold – A feature that allows the phone user to put one call on hold while a second call is made or answered.

call set-up time – The time required to screen dialed digits, determine the proper routing for that call, and then send the call on its way.

call screening – Extra processing on a call that allows the user to handle a call in a particular manner. Call screening allows users to restrict calls to specific areas, time of day, day of week, etc. This is particularly relevant for 800 and virtual network services.

call waiting – An optional feature that provides audible or visible cues to let a telephone user know that he or she has another incoming call. The user can put the existing call on hold or hang up to answer the waiting call.

Caller Groups – Part of AT&T's call screening option for SDN. These are groups of SDN stations or user authorization codes that must be defined so call privileges may be assigned to them.

carrier – In this book, used to refer to a long distance provider (interexchange company).

carrier access code (CAC) – See CIC.

carrier identification code (CIC) – See CIC.

casual caller – A long distance term referring to a user that dialed the 10XXX equal access code in order to reach a long distance carrier, where XXX is the carrier identification code of the long distance carrier.

CCIS (common channel interoffice signaling) – A network control and call set-up system that employs digital transmission of information between central offices for a group of trunks on a separate, dedicated channel.

CCITT (Comite Consultatif International Telegraphique et Telephonique) – A consultative committee to the International Telecommunication Union (ITU) to regulate and recommend international standards for satellite, telephone and telegraph services.

CEI (comparable efficient interconnection) – A standardization proposal which permits all like enhanced service vendors to buy basic network features, interfaces and services on an unbundled basis from the local Bell operating company. Think of CEI as equal access specific to one particular type of enhanced service. (See also BOC, BSE, BSA, ONA.)

Cellular Geographic Service Area (CGSA) – The legally authorized operating area for a cellular telephone company.

centralized database – This is a central information processing facility where all customer and network records can be entered, maintained, and made available for access.

central office (CO) – A facility where calls are switched. Each central office has an "address" that consists of the area code and first three digits of a phone number. A serving central office is the central office that is responsible for access and egress for the customer. Also called an "exchange."

Centrex – A switch based local telephone company service that takes advantage of local central office intelligence to compete against office PBXs and key systems. Centrex uses the power of the central office switching intelligence to route internal calls from one station to another, to direct-dial outbound calls, and to mimic many PBX-like service features. Centrex is ideal for small businesses that need to grow quickly.

channel – A functional path for the transmission of information. Also, one circuit within a multicircuit transmission facility.

channel bank – A transmission facility used to multiplex and demultiplex several different channels on a frequency- or time-division basis. A channel bank can thus convert analog voice-grade signals into digital form and the reverse.

chargeback – In 900 services, when the end user pays the LEC bill with line-item exceptions. The LEC will charge that portion of the bill back to the 900 service provider.

CIC (carrier identification code) – A three-digit number assigned by Bellcore that is unique to a long distance carrier. This code is used in the public network to identify the carrier's calls. With equal access, a user calling through casual access must dial 1 + 0 + the three digit CIC of the carrier wanted, plus the desired telephone number. The local exchange company will route the call to the carrier's POP. The CIC is a nationwide code. It is also called the carrier access code.

CIMS – See Customer Information Management System.

circuit – The physical path for transmission of information between two points.

CLASS – See custom local area signaling service.

class 5 office – Terminology for the lowest level central office, on a scale of classes ranging from 1 to 5. Class 5 offices serve as the network entry point for user access lines. They also serve as a switching point for local calls. Also referred to as "End Offices."

clock – Equipment designed specifically to give computer or communications systems time reference points. Clocks are required for systems that depend on synchronization for functionality.

collection – The collection of revenues from end users. Usually provided with billing service. (see billing).

common carrier – A private company, regulated by state and federal agencies, that provides telecommunication services or communications facilities to the general public.

common channel signaling – See CCIS.

conditioning – The ability to enhance the transmission quality of a circuit by changing its electromagnetic properties for a specific set of performance characteristics.

Consent Decree – Generally refers to the 1956 agreement that settled a federal antitrust suit, initiated in 1949, as modified by the AT&T divestiture agreement of 1984.

CPE (customer premises equipment) – Telecommunications terminal equipment at the customer's location. Such equipment can be as simple as a phone, or as complex as multiline telecommunications equipment including PBXs and key systems.

crash – When a computer or communications system fails unexpectedly. Most of the time, the failure is complete.

crosstalk – The sound of two simultaneous conversations on a telephone line.

custom calling – A group of options or enhancements, that offer the some of the sophisticated features of PBXs to smaller residential and business customers. Examples of custom calling would be call waiting or speed dialing. (See also custom local area signaling service.)

Customer Information Management System (CIMS) – An MCI Vnet feature, this allows the customer to create, view, and revise Vnet data, including: 1) changing dialing plans; 2) changing direct termination overflow; 3) assigning range privileges; and 4) activating and deactivating ID codes.

custom local area signaling service (CLASS) – A set of local calling enhancements to basic local exchange telephone service. CLASS features use digital switching and out-of-band signaling to provide extra processing on calls, including the ability to trace, redial, reject, or forward calls. (See also Custom Calling, out-of-band signaling, Signaling System 7.)

customer premises equipment – See CPE.

Customized Range Privilege – A feature under MCI's Vnet service in which a customer can impose 1-89 restrictions on calling privileges. Range privileges can be specified by NPA, NPA/NXX and country code.

D channel – The "Data" channel used in signaling and data transmission as outlined in ISDN standards. The D channel uses packet-switched channel technology to send and receive network control signals for establishing the paths and parameters of an ISDN call. The D channel is also capable of carrying data into the ISDN terminal, including information about the call itself from the network or related data sent from another terminal device. (See also B channel, basic rate interface, ISDN, primary rate interface.)

DAL – See dedicated access line.

Database – A collection of indexed, filed or otherwise organized data that can be accessed by a network for retrieval of information or updating.

dataphone digital service (DDS) – A dedicated digital data transmission service using digital private lines for synchronous transmission of data at speeds of 2.4, 4.8, 9.6, or 56 kbps between subscriber locations.

DDS – See dataphone digital service.

deadspot – An area within a cellular system where service is not available due to terrain, electronic interference or foliage.

dedicated access line (DAL) – A nonswitched access line that connects an inter-LATA or interstate long distance user directly to his long distance carrier's point of presence. Because the circuit runs directly to the carrier POP, the local exchange carrier is not able to screen calls that pass through the local network, as with WALs. (See also equal access, WAL.)

dedicated line – A point-to-point communications channel that is reserved exclusively for the use of a particular subscriber. This term can refer to an access link to a long distance carrier POP, in which case it is a dedicated access line (DAL) , or it may refer to a long haul circuit, in which case it is a private line. (See also DAL, private line.)

deregulation – Change in the status of a regulated industry, such as the telecommunication sector, whereby all or many of the rules and statures governing its behavior are removed.

D-4 framing – The most popular framing format for T-1 is D-4 framing. This name is derived from AT&T's D-series of channel banks or D-channels in which there are 12 separate 193-bit frames which constitute a superframe. The superframe format has been used by channel banks for more than 20 years. Inbound analog voice signals are digitized and coded into a standard T-1 format by sampling them 8,000 times a second. This analog signal is coded into 8-bit words or a byte, to form the fundamental T-1 building block of 64 kbps for voice. This is digital signal level 0, or DS-0, and 24 DS-0s are combined into a serial bit stream using Time Division Multiplexing (TDM) to create T-1 frames. The frame equals 24 x 8 bits or 192 bits. To decode and direct each byte of information, the receiving terminal must to be able to tell what information is carried on which T-1 channel. This is provided by a synchronization bit, called the framing bit. Bringing the overall length of the T-1 frame to 193 bits. Transmission is at a rate of 8,000 frames per second x 193 rate, or 1.544 Mbps. This is the DS-1 rate. (See also digital hierarchy, ESF)

dialed number conversion – A virtual network feature that enables the network to convert seven-digit numbers to ten-digit numbers, and vice versa.

dialed number identification (DNIS) – Used with 800 services, enables a customer to receive calls from several 800 numbers over the same access facilities but still identify which 800 number was dialed by the caller.

dial 1 – A generic name for dialing "1" plus the telephone number to complete a long distance call. (See also equal access.)

Dial-1 WATS – The trade name for US Sprint's service designed for small business customers.

dial-up – Local services in which a user dials a local number and then a code number to gain access.

DID – See direct inward dialing.

digital – Describes method of storing, transmitting, and processing information. The method employs the use of distinct electronic or optical pulses which represent the binary digits (bits) 0 and 1. Digital transmission and switching facilities use a sequence of discrete, individual pulses to represent information, as opposed to the continuous, variable signal of analog technologies. (See also analog.)

digital hierarchy – A standardized hierarchy of increments for multiplexing digital channels into a T-carrier or other type of carrier system. The lowest rung of the hierarchy is digital signal level zero (DS-0) – a single 64 kbps digital voice channel. The next level is DS-1, which incorporates 24 DS-0 channels into one 1.544 Mbps signal. Four DS-1 signals are multiplexed to form one DS-2 signal of 6.312 Mbps. Seven DS-2 signals or 28 DS-1 signals form one DS-3 signal or 44.6 Mbps. One DS-3 signal carries 672 digital voice channels. Other steps in the hierarchy that are commonly found are DS-1C (48 voice channels multiplexed into a 90 Mbps signal) and DS-3C (two DS-3s multiplexed into a 90 Mbps signal). At the very top of the hierarchy is DS-4 (274 Mbps, equal to 168 DS-1s). The higher levels are generally used in interoffice transmissions. This hierarchy is applicable for North America only, other parts of the world use different hierarchies and transmission increments.

Digital Routing – AT&T's switched 56 kbps data transmission service for SDN. There are two forms of access: 1) static, or full-time access; and 2) dynamic, or on-demand, access through a special feature code. Static access requires dedicated T-1 or DDS facilities. (see also Accunet Switched Data Services.)

digital switch – A piece of equipment used to set up paths between users for transmission of digital signals.

digital switched access (DSA) – Feature group D switched access in the local network that is carried over all-digital facilities. (See feature group D)

direct distance dialing (DDD) – The ability of the public switched telephone network to automatically route calls to an area code, exchange prefix and subscriber line number – without the help of an operator. Prior to DDD, all long distance calls were operator assisted. A variant is IDDD, international direct distance dialing.

direct inward dialing (DID) – A feature of PBXs and Centrex systems which allows callers on the public switched network to dial directly a specific PBX or Centrex station.

Direct Termination Overflow – MCI's public switched network overflow feature for Vnet.

directory assistance (DA) – A service that provides the listed telephone directory number for a requested person or company.

distributed database – An information system in which the data are dispersed among many points within the network instead of being concentrated at one central location. This is opposed to centralized database structures. (See also centralized database.)

divestiture – In this book, an agreement, known as the Modification of Final Judgement, implemented on Jan. 1, 1984, that caused AT&T to divest itself of its Bell operating companies. The divestiture agreement settled a 1974 federal antitrust case against AT&T. Judge Harold Greene in the U.S. District Court for the District of Columbia gave final approval to the agreement and presides over its implementation. The agreement was presented as a modification of the 1956 Consent Decree that ended a 1949 antitrust suit against AT&T. This is also referred to as the Consent Decree. (See also AT&T, Bell System, BOC, Consent Decree.)

DNIS – See dialed number identification.

DSA – See digital switched access.

DS-0 – Digital signal level 0, a digital 64 kbps transmission channel. This is the lowest level of the digital multiplexing hierarchy. (See also digital hierarchy, DS-1, DS-3).

DS-1 – Digital signal level 1, a 1.544 Mbps digital signal comprised of 24 multiplexed DS-0 64 kbps digital channels carried over a T-1 facility. (See also DS-0, T-1, digital hierarchy.)

DS-3 – Digital signal level 3, a 44.6 Mbps digital signal comprised of 28 multiplexed DS-1 signals carried over fiber optic lines at a transmission speed of 44.6 Mbps. (See also DS-1, T-3, digital hierarchy.)

DTMF – See dual-tone multifrequency signaling.

dual-tone multifrequency signaling (DTMF) – A technology that uses specific combinations of two different electronic tones to represent dialed digits. Also referred to as Touch-Tone.

EDI – See electronic document interchange.

800 Service – A long distance service designed to charge the telephone call to the terminating, rather than the originating party. Because the caller does not pay, the service is often (inaccurately) called "toll-free." The service takes its proper name from the special access code (SAC) – 800 – which it uses to signal the local serving central office to reverse the charges.

egress – The opposite of access. Egress refers to exiting a network.

electronic document interchange (EDI) – Data communications technology that is being used to replace paper-based business transactions, including long distance bills, by sending the information directly from one computer to another.

electronic serial number (ESN) – The unique identifying number of a cellular telephone. The ESN is embedded in the telephone's circuits by the manufacturer.

electronic switching system (ESS) – A type of central office switch sold by AT&T's equipment division. The switches are controlled by stored programs, and form the foundation of AT&T's intelligent network.

electronic tandem network (ETN) – A network designed send traffic through two or more relay points which are connected in tandem. (See also tandem office.)

embedded base – A vendor's or carrier's presently installed customers.

equal access – The ability of a customer to select, by the digits that he or she dials which long distance provider will carry the call; or the ability of each long distance carrier to obtain connections from the local exchange telephone companies of equal quality to those of AT&T.

ESF (enhanced superframe format) – A framing format for digital T-1 transmission in which 24 frames, instead of the 12 frames stipulated under the D-4 Superframe format, are grouped together. The 8,000 bps organization is segmented into 2,000 bps for network information, 4,000 bps for error checking, and 2,000 for synchronization. The new 24-frame format allows for additional network signaling information to provided better diagnostics and network information during transmission. (See D-4.)

ESP (enhanced service provider) – A company that adds special services to voice and data messages before transmitting them. An example would be a company that compiles data results before transmitting the data to its final destination.

ESS – See electronic switching system.

ETN – See electronic tandem network.

exchange – See local exchange.

Expanded Service Management System (ESMS) – Part of AT&T's Unified Network Management Architecture (UNMA). This SDN feature allows customer to access three elements: network configuration (software database definition), access (alarm and testing) and costs management elements, which include: 1) next day on-line traffic analysis, both daily and weekly; 2) management reports form call detail; 3) bill or call detail on magnetic tape; 4) customized decentralized billing.

extended digit screening – A capability of a network to screen all ten digits of a telephone number as it passes through the switch.

facilities – The physical components that make up a private telephone service.

FCC (Federal Communications Commission) – The U.S. government agency which regulates radio and TV broadcasting and interstate and international telecommunications.

feature group – A specific type of local exchange switched access which is available to interexchange carriers and users. There are four feature groups – A, B, C, and D.

feature group A (FGA) – A line-side connection for originating and terminating traffic. To access a Feature Group A circuit, customers must dial a seven-digit local number to reach an interexchange carrier, and enter an identification code plus the termination phone number. (See also access charge, line-side connection.)

feature group B (FGB) – A trunk-side switched access connection for originating and terminating traffic, providing superior transmission quality to Feature Group A circuits. Feature Group B circuits are accessed through dialing a 950-number and the telephone number. (See also access charge, trunk-side connection.)

feature group C (FGC) – A trunk-side switched access connection directly linking local exchange company network with AT&T. Feature Group C access is available only to AT&T; all other carriers must acquire the comparable Feature Group D circuits. Both offer the highest quality of transmission, as well as complete series of access features. Users dial "1" plus the telephone number to complete the call. (See also access charge, trunk-side connection.)

feature group D (FGD) – A trunk-side switched access connection that represents equal access. Comparable to the feature group C connection. Customers in exchanges where Feature Group D is available can presubscribe to any one long distance company. Users dial "1" plus the telephone number to complete the call.

Equal access customers may access other long distance companies by dialing their 10XXX equal access code. (See also access charge, carrier identification code, switched access, trunk-side connection.)

fiber optics – A technology that uses filaments of glass or other transparent materials usually in a wire or cable as the transmission medium for light pulses that represent data, image and sound. Because fiber optic technology employs light as the transmitting vehicle, it offers extremely high transmission speeds.

flat rate – Pricing in which the price charged per billing period is the same, regardless of usage.

Flexible Routing – This SDN feature allows the alternate routing of traffic terminating at a special access on-net station or group of stations to any one of the following: 1) another on-net station; 2) a group of on-net stations; 3) an intercept announcement; 4) an off-net station or group of stations.

Follow Me Roaming – A service provided by GTE Mobilnet that allows a cellular user to receive calls made to his cellular telephone when traveling in another city.

FONCARD – US Sprint network remote access option in which a user dials 800-887-8000 and then 0 plus a ten-digit number, plus a fourteen-digit code.

forced on-net calling – A virtual network feature which provides an automatic ten-digit to seven-digit translation function that enables a company to retain ten-digit calling plans for calls made on the network.

foreign exchange (FX) – A service marketed by long distance and local exchange carriers in conjunctions that combines technologies of the two to offer origination or termination of calls in a calling area, to or from a distant calling area. A private line connects the subscriber with a central office or tandem switch outside the local calling area. Under FX service, the phone number in the "foreign" area is now part of the subscriber's local service. FX service is typically used by companies that have a large amount of calling from another area. Airline reservation lines are prime examples.

four-wire transmission – A transmission of two-way signals that uses separate paths to transmit and receive.

fractional T-1 – Private line service that offers digital capacity in fractions of T-1 circuits, for a portion of the cost of T-1. For example, a user could purchase half a T-1 for 60 percent of the cost of a full T-1 circuit.

FX – See foreign exchange.

GAB – Stands for "Group Access Bridging". Allows 900 carriers to link several lines at once, permitting group calling sessions.

gateway – An interface where otherwise incompatible networks or devices meet and interconnect. Particularly useful where domestic networks interconnect with those of foreign carriers.

Gbps – Gigabits per second, or billions of bits per second. (See also bps.)

GLOBAL FON – US Sprint's marketing alliance with Cable and Wireless Ltd. for the joint marketing of international long distance programs, notably between the U.S., the U.K., and Hong Kong.

Global Value Plus Program – An MCI international program that gives users an extra discount on customer-dialed calls to all IDDD countries, for all dedicated and switched Vnet calls.

Greene, Harold – The federal judge of the U.S. District Court for the District of Columbia who is responsible for judicial enforcement of the terms of the AT&T divestiture agreement, which led to the breakup of the Bell System. Judge Greene is also charged with the responsibility of ruling on requests for waivers or modifications of the terms of the divestiture agreement. (See also divestiture, Bell System, Consent Decree.)

handoff – In cellular communications, the transfer of a conversation from one voice channel to another.

home carrier – The cellular telephone company with which a cellular telephone is registered and to which a user pays monthly access charges.

host carrier – The cellular telephone company providing service to a subscriber from a "foreign" carrier. (See home carrier.)

hybrid – The combination of two or more systems or technologies. Particularly relevant to virtual networks combining private and public networks – i.e., private line-virtual network hybrid networks.

Improved Mobile Telephone Service (IMTS) – A mobile telephone service that is presently available in more than 2000 U.S. and Canadian cities which is not compatible with cellular service.

in-band signaling – A method of signalling which allows the network control and set-up of a call on the same channel as used by the call itself. In-band call set-up uses the communications channel to establish call parameters. The transmission of the call itself must wait until set-up is finished.

Information Provider (IP) – The industry term for the sponsor of a 900 service. This company provides the content to an information service, either in voice or text formats.

Integrated Network Management System (INMS) – MCI's future network management system. A user will be able to manage: 1) operations; 2) trouble; 3) configuration; 4) performance and planning; 5) billing and costs; and, 6) order entry and tracking.

Integrated Network System Interface and Terminal Equipment (INSITE) – Sprint's future VPN network management system. A user will be able to: 1) review direct connection data; 2) activate and deactivate authorization and accounting codes; 3) trouble reporting entry and tracking; 4) manage 700, 800, and 900 services; 5) manage calling cards; 6) retrieval of archived billing data; 7) retrieve network performance data; and 8) perform customer-controlled reconfiguration. First phase (of four) is available now.

installed base – See embedded base.

integrated services digital network (ISDN) – A switched digital transmission network allowing end-to-end digital connectivity for simultaneous transmission of voice and data. More than anything else, ISDN is a series of standards for this digital connectivity. ISDN uses out-of-band signaling protocols for transmission. Through multiplexing, ISDN users can access multiple circuit-switched and packet-switched channels through a single terminal that combines voice and data features. (See also basic rate, primary rate)

Interactive – A term which applies to those 900 services that can accept DTMF signaling (touch tone) from the caller. This allows the user to control his interaction with the 900 service program.

interconnection – A connection of telephone equipment to a telephone network, or the connection of one communications carrier to another.

interexchange carrier (IXC) – A long distance carrier which has obtained authorization from the FCC to offer long distance communications services between interstate LATAs, or a carrier which has obtained the necessary authorization by a state's public utility commission to provide intrastate long distance communications services.

interface – 1) The point of linkage between two or more devices that allows those pieces of equipment to interact. 2) Standards and protocols which allow dissimilar devices to interconnect and communicate.

inter-LATA – Communication between different local access and transport areas. (See also local access and transport area.)

international VPN – US Sprint's international virtual network service that will be integrated with its domestic VPN service. Countries served include the United Kingdom and Hong Kong. Part of the Global FON program. (see Global FON.)

internetwork calling – A feature of virtual networks that helps users interconnect their virtual network with other virtual or private networks.

intra-LATA – Communications within a given local access and transport area. (See also local access and transport area.)

ISDN – See integrated services digital network.

IXC – See interexchange carrier.

kbps – Kilobits per second, or thousands of bits per second. (See also bps.)

key telephone system – A telephone system capable of handling multiple lines, designed for smaller business for use as their primary telephone system. They are also used by large businesses to form an extension to their larger main phone system. Key systems have limited options – outgoing lines are selected by pressing a button on the telephone set.

LAN – See local area network.

LATA – See local access and transport area.

leased line – Voice or data circuits which are leased from the local telephone company or the interexchange carrier in order to connect two or more of a user's locations for his exclusive use. Also known as a private line.

LEC (local exchange carrier) – A telephone company that provides local service and intraLATA toll service.

LIDB – See line information database.

line information database (LIDB) – A cross-referencing database that is under development by the Bell operating companies to support database-intense applications such as 800 number translation and credit card processing in response to an order from Judge Greene to provide advanced 800 service capabilities to all long distance carriers. The database will use Signaling System 7 protocols for access.

line-side connection – A connection that links between a circuit and a central office switch port that treats the circuit as a virtual circuit or local line connected to a telephone. As such, the connection offers transmission and signaling features usually afforded a connection between a central office and a standard telephone. This is opposed to trunk-side connections. (See also line, feature groups, access charges.)

local access and transport area (LATA) – One of 161 geographical areas that were created at divestiture to differentiate local phone service from long distance service.

local area network (LAN) – A network which links a variety of private data communications devices, including computer terminals, micro- and minicomputers, printers and telephones. LANs vary in terms of size, complexity, and geographical coverage.

local central office – The switching office that provides the initial screening and switching on calls coming from a subscribers' local loop. Also provides similar functions for calls bound for the subscriber's local loop.

local channel – Generally, same as local loop. However, increasingly, this refers to the portion of an inter-LATA or interstate WATS service or private line service provided by a local telephone company to link the customer's premises with the interexchange carrier's point of presence.

local exchange – Also called a central office, this is a switching control center which serves telephone lines that are usually identified by the area code and exchange digits of a phone number.

local loop – A communication link between a customer's premises and his serving central office capable of originating or terminating both local and long distance traffic. Also referred to as the subscriber loop. (See also equal access.)

local measured service (LMS) – See measured service.

Location Sharing – An SDN feature that permits any on-net station, served by special access facilities, to be identified as an on-net station for more than one SDN.

long-haul – Same as long distance communications. This term typically refers to any inter-LATA circuits – intra- or interstate.

Mbps – Megabits per second, or millions of bits per second. (See also bps.)

MCI – One of the "big three" in long distance telecommunications services. MCI is the second largest company, with a market share of approximately 11 percent.

MCI FAX – MCI's facsimile network offering that provides enhanced facsimile features.

MCI Vnet Card – MCI's network remote access option in which a user dials a customer-specific MCI Vnet 800-number, enters either a "0" plus a ten-digit number plus a fourteen-digit code, or "8" plus a seven-digit on-net number plus the fourteen-digit code. Cards can be restricted by customer-defined range privileges.

measured service – Local exchange service billed on a usage-sensitive basis, instead of on a flat rate. Local calls are priced on many of the same criteria that long distance calls are made – distance and length of call and the time of day. (See also flat-rate.)

Meeting Channel – The trade name for US Sprint's videoteleconferencing network. The service provides two-way, full-motion, color video conferencing.

Megacom Service – The trade name for an AT&T switched service which offers discounts to customers that have an extremely high volume of outbound WATS-type traffic. Its rate schedules resemble AT&T's SDN Schedule B rates. Distance is calculated according to a mileage band system. Timing is in 18-second initial period, and subsequential 6-second increments. There are three time-of-day rate categories – day, evening, and night and weekend. Megacom users typically use T-1 access circuits. International termination capability was added in 1990. This is in the 30 second initial and 6-second subsequent pricing format and uses rates formerly part of the World Connect program.

message rate – Local exchange service in which a user is charged a flat rate for each call. The charges are commonly called "message units." Message rate is different from both measured service and flat-rate monthly service. (See also flat rate, measured service.)

message unit – See message rate.

MHz – Megahertz, or millions of cycles per second.

minimum average time requirement – A threshold criterion occasionally used in telecommunications tariffs. It equals the total number of minutes of calling per month divided by the total number of calls in a month.

MTS (message telecommunications or toll service) – Basic switched long distance telephone service designed primarily for transmission of voice communications. MTS lines can also carry data at slow speeds (up to 4.8 kbps, sometimes up to 9.6 kbps) without special conditioning of the lines.

multiplexed channel – A communications channel used by multiple users or devices at the same time.

multiplexer – A device that multiplexes communications signals. (See also multiplexing.)

multiplexing – An process that merges a large number of lower-speed information streams into one higher-speed circuit. There are a number of different multiplexing formats. Among the most popular are 1) frequency division: splitting the total available bandwidth of the high-speed line into narrower bands, 2) time division: splitting a single common channel among several different transmitting devices, each of which is given a specified time slot in which to send information over the channel. 3) statistical time division: splitting a common channel among several different transmitting devices, each of which is given access to the channel for data transmission based on the amount of information each device has to send.

multidrop line – A communications line, or circuit, which employs multiple stations and multiple locations.

mux – Short for multiplexer.

N – Standard reference for numbering plans. N stands for any number between 2 and 9.

NANP – See North American Numbering Plan.

narrowband – A communications channel with a bandwidth less than that of a voice grade channel.

network – A system in which a number of communication paths are connected by communications channels. These communications paths may carry voice, data, facsimile, and video messages. Typically, there are private networks and public networks – the private reserved exclusively for the use of its organization, and public available to all who wish to use it under its terms and conditions.

network architecture – The configuration of the network as a system through the design and arrangement of the interaction and performance of its protocols and its functional and logical components.

network channel terminating equipment (NCTE) – A device installed at the customer's premises which acts as the interface for signal-conditioning and channel-testing functions of digital private lines. T-1 services typically need NCTE, as do other digital services depending on the application and service.

network control signaling – Signal transmissions in a network the function of which is to manage, monitor, and inform the network, supervise all circuits, perform call path set-up, communicate between switches, and signal the switches.

Network Information Management System (NIMS) – This MCI Vnet feature provides on-line access to next day traffic information and near real-time access to operational and performance information. Both hourly and daily reporting are kept for the current day and the previous seven days. Common uses are: 1) performance reports; 2) trouble detection; 3) trunk measurements; and 4) circuit status reports.

network intercept announcements – With virtual networks, these allow the customer to have calls intercepted by prerecorded announcements to relate calling restrictions or network conditions.

network interface – Any sort of physical connecting point between two networks, or between a customers premises and a network.

network management – Procedures, and operations through equipment and software, intended to maintain nearly normal network operating conditions in the face of unusual demands or emergency situations that would otherwise result in slowing down of traffic or service failures.

Network Remote Access – AT&T SDN remote access feature that has three options. Under option 1 (Customer-Administered), companies are assigned an 800 number for each remote access option ordered. Authorization codes are assigned and maintained by the customer. Under option 2 (PLUS), companies are afforded the same privileges, as with option 1, but authorization codes are assigned and managed by AT&T. In addition, the customer has the ability to dial 0 plus the called number and enter the authorization code – no operator assistance is required.

Under option 3 (Express), users have a customer-specific 800 number that terminates directly to one on-net station on the SDN. (See also remote access.)

NIMS – See Network Information Management System.

nonwireline – A company operating the A block of cellular channels.

North American Numbering Plan (NANP) – The standard ten-digit numbering plan used in the public switched telephone networks of the U.S., Canada, northern Mexico and the Caribbean. The NANP divides those areas into geographic regions, which are assigned their own unique three-digit codes, commonly called area codes. The NANP also reserves certain phone number prefixes for specific purposes, such as the 700, 800, and 900 prefixes. Under NANP there are 144 possible area code combinations. Of the 144 possible area codes, 137 are in use and with two more will be in use by the end of 1991. (See also numbering plan area, service access code.)

numbering plan – An arrangement of numbers in which each terminal or station on a telecommunications network is identified by its own number.

Numbering Plan Area (NPA) – A geographic area identified in the North American Numbering Plan by a unique three-digit area code. (See also North American Numbering Plan.)

NYNEX – One of the seven regional holding companies created from at divestiture. NYNEX owns and controls New England Telephone and New York Telephone.

O plus (O+) – Commonly called operator-assisted or credit card calls. The calling party dials O plus a seven-digit or ten-digit phone number. In most cases, the caller receives a tone prompting input of a credit card number. The caller enters the number of his calling card after a tone, bypassing the operator. If it is an operator-assisted call, the caller stays on the line and an operator answers. In some cases, the operator answers regardless, due to the switching equipment in place.

off-hook – The electrical signal sent upon lifting the handset of a telephone whereby the line is ready to receive or send a communication and is closed (busy) to other signals while doing so. The central office switch reads it as either a response to a ring signal or a request for dial tone.

Off-Network Overflow on Termination Busy – AT&T's public switched network overflow feature.

ONA – See Open Network Architecture.

on-hook – When handset is in its cradle and no electrical signal is being sent from the phone to the central office. The line is inactive by open to communication upon receipt of a ring signal or request for dial tone.

on-line – The placement of a terminal or device in the active or connected state whereby an element or function is part of the circuit, loop, or network.

Open Network Architecture (ONA) – An industry design for packaging exchange company basic network facilities and services that standardizes the means which vendors of basic or enhanced services can buy basic network functions and interfaces on an unbundled equal-access basis. (See also BSE, CEI, Computer Inquiry, enhanced services.)

operating company – Generally, a local exchange company including the 22 telephone companies formerly part of the Bell system and independents such as Southern New England Telephone. (See also Bell operating company.)

originating call identification – An intelligent network feature that allows a customer to receive the telephone number of an incoming call with the call itself.

out-of-band signaling – The transmission of network control signals over a channel different from the channel used for the transmission of the user message itself. This is done by segmenting the circuit's bandwidth into two portions – one for signaling and one for the voice or other form of communication. (See also CCS, in-band signaling, network control signaling.)

PABX (private automated branch exchange) – See private branch exchange.

Pacific Telesis – One of the seven regional holding companies created at divestiture. It is the holding company for the Pacific Bell and Nevada Bell telephone companies.

packet switching – A data transmission method that sends bundles of data, usually in binary form, through a network to a terminating location. Each bundle, called a packet, is assigned a specific identification number and termination address. In transmission, the packets are sent by the most efficient means of transport at the particular time; packets of related data may go over entirely different routes to get to their destination. At the termination point, the packets are collected and reassembled in their correct order.

Partitioned Data Base Management – An SDN feature that allows the customer to divide an SDN into multiple smaller SDNs. Three possible billing arrangements exist: 1) One bill for the SDN, regardless of the number of subnetworks, 2) One bill for each subnetwork, or 3) Each location receives its own bill via multilocation billing.

passive – A term which applies to those 900 services that do not allow any interaction with the caller. The person dialing into the 900 program can only listen to the program.

PBX – See private branch exchange.

Plain Old Telephone Service (POTS) – Basic telephone service for voice communications.

point of presence (POP) – An interexchange carrier's switch in a LATA where the long distance carrier lines connect with the local exchange companies' lines within that LATA. Typically, a carrier has one POP per LATA, but in areas where demand is high, there may be more than one point of presence.

Polling – A 900 application that allows callers to dial into one of two or more passive 900 numbers or one interactive 900 number to quickly express a view on an issue.

POP – See point of presence.

port – A physical interface point on a computer or communication device to which peripheral equipment or communications lines are connected.

POTS – See Plain Old Telephone Service.

PRI – See primary rate interface.

price cap regulation – A method of regulation recently approved to govern AT&T's earnings in the marketplace. Price caps replace rate-of-return regulation. With price caps, the carrier is given a price ceiling for its various telecommunications services. The regulated company is free to lower rates to any point below the ceiling level without prior approval.

primary rate interface (PRI) – The main interface between a customer's premises and carrier networks for ISDN business services. It is the high-volume interface, covering the connection of an ISDN PBX or other large business communications system to a 1.544 Mbps digital facility (North America) or a 2.048 Mbps digital facility (Europe). In North America, the PRI 1.544 Mbps has 23 message-bearing 64 kbps B channels for voice, data, etc., transmission, plus a 64 kbps D channel for network signaling. (See also B channel, basic rate interface, D channel, integrated service digital network.)

Prism – MCI's WATS family. Includes Prism I, MCI's highest volume service.

private branch exchange (PBX) – A user's telephone switch, located on its premises, which handles all outgoing and incoming communications. The user can customize the applications within the switch to provide a range of options and features, including enhanced call screening and routing functions. Also referred to as PABX.

private line – A nonswitched telecommunications channel that provides full-time dedicated communications between two points. (See also dedicated line.)

private network – A network that is designed and maintained for the expressed use of an internal group of users. This is opposed to a public network.

Private Network Interface – An SDN feature that allows interconnection with private networks, while maintaining a uniform dialing plan.

PSTN – See public switched telephone network.

PTT (Postal, Telegraph and Telephone agency) – A national government's telecommunications authority or agency responsible for the design, operation and interconnection of its domestic telephone system.

public switched network overflow – A virtual network feature that automatically completes calls over the public switched telephone network when terminating dedicated facilities are busy or down.

public switched telephone network (PSTN) – The equipment and switching intelligence that provides routing, switching and transmission among public users. Generally, this term refers to the entire switching network consisting of local and long distance switched service networks.

rate – In pricing, the amount that a user is charged for a particular service. For regulated carriers, these prices are usually made public through tariffs filed with the appropriate regulatory bodies.

rate center – A geographic location that serves as a reference point for calculating the distance of telephone calls for rating purposes. Each rate center is fixed by vertical and horizontal coordinates, similar to longitude and latitude map coordinates.

rate of return – A form of regulation that limits the amount of profits that a regulated carrier may earn, based on the percentage of net profit relative to capital investment.

RBOC – See regional Bell operating company.

redundancy – Incorporation of duplicate components into a system so that a duplicate component immediately takes over if the primary component fails.

regional Bell operating company (RBOC) – One of seven regional companies created by the AT&T divestiture to take over ownership of the Bell operating companies within their region. They are Ameritech, Bell Atlantic, Bell South, NYNEX, Pacific Telesis, Southwestern Bell and US West. The RBOCs also have set up numerous unregulated subsidiaries engaged in a variety of communications-related and noncommunications businesses. The divestiture agreement RBOCs from engaging in certain business activities, such as providing long distance service, but includes mechanisms for review, waiver, modification or removal of the prohibitions. The RBOCs are also known as regional holding companies (RHCs). (See also operating company and the individual RBOC names.)

regional holding company – See regional Bell operating company.

remote access – 1) The ability of a long distance service (usually virtual network) user to access his network by dialing a number, entering an authorization code (if required), and then placing the call. 2) A similar feature on most PBXs that enables a user at an outside location to dial into a corporation's PBX to access certain features, including completing calls using the companies WATS, MTS, or private line services.

remote call forwarding – A service feature that lets users list a number in the central office. Calls coming to this remote call forwarding number are automatically forwarded to a designated answering location.

remote station – A station or terminal which is physically remote from a main station or computer but that can gain access through communications circuits.

resale carrier – A carrier that obtains communications services from another carrier for resale to the public for profit. Typically, these carriers do not own their own facilities.

reseller – See resale carrier.

RFP – See request for proposals.

RHC – See regional Bell operating company.

roamer access number – In cellular communications, name given to a public ten-digit number that allows telephone users to receive calls when they are outside their home service areas. A person wishing to contact a cellular subscriber dials the roamer access number where the subscriber is located, waits to hear a tone, then dials the area code and number of the cellular telephone.

roaming agreement – A mutual agreement between two cellular telephone carriers to accept the calls of each other's customers. A customer is billed by the home carrier for any calls placed on the host carrier plus any applicable service fees.

RoamingAmerica – In cellular communications, a service provided by Appex that allows a cellular user to receive calls made to the user's cellular telephone when in another city.

Route Advance – US Sprint's public switched network overflow feature for VPN.

SAC – See service access code.

SAL – See special access line.

SCP – See service control point.

Screening Groups – In AT&T's SDN service, these are the definition sets for calling privileges under its call screening option. Screening groups may be defined as: 1) a listing of specific on-net addresses; 2) all exchanges within an area code; 3) a listing of specific exchanges in an area code; or, 4) all off-net addresses.

SDN – See Software Defined Network.

SDDN – See Software Defined Data Network.

secondary channel service – A digital data service (DDS) option that provides users with a network interface capable of accommodating a second channel derived from the primary channel by means of customer provided multiplexing devices. Secondary channel services are typically employed by users for monitoring a network without disturbing data transmissions and for encryption of data during transmission. (See also DDS.)

service access code (SAC) – Code of N00 (N denotes any digit from 2 through 9) which replaces the usual area code in phone numbers for special network uses.

Three of the eight possible service access codes have been assigned: 900 for direct-dialed information services, 800 for toll-free dialing services, and 700 for direct-dial access to specialty services provided by interexchange carriers.

service area – The geographic area served by a telecommunications provider.

service aggregator – A form of third party marketer that became popular in 1990. An aggregator subscribes user firms to its long distance service to gain higher discounts with its long distance provider. The higher discounts are usually split between the aggregator and the user firm.

service control point (SCP) – In Signaling System 7, a service control point is a computer database that stores information for such network activities as credit card validation or translation of an 800 number into a regular ten-digit phone number. (See also signal and Signaling System 7.)

serving wire center – See wire center.

Seven-Digit International Dialing – An SDN feature that allows users to define a specific international number that will appear as a seven-digit on-net number.

Signaling System 6 – A high-level signaling protocol used by AT&T today. SS6 allows for fast processing of calls and this has been a decided edge for AT&T in competing against MCI and US Sprint. The carrier is migrating to SS7. (See also Signaling System 7.)

Signaling System 7 – A specific network control system comprised of protocols for the interpretation and use of a repertoire of network control and operation signals, and the associated hardware and software needed for transmission, reception and interpretation of the signals. Signaling System 7, as the latest-generation network signaling system, accommodates enhanced 800 service, wide-area Centrex services, virtual private networks, and other types of advanced telecommunications services. Signaling System 7 also conforms to the international signaling protocols established by the CCITT for integrated services digital networks (ISDN). Signaling System 7 employs its own network, separated from the network that carries customer messages. It is now being implemented in the networks of most major local telephone companies and long distance companies. In handling advanced network services in an SS7 signaling network, the switch at the originating end uses a packet switch called a signal transfer point to send a query to a network information database called a service control point. In response to the specific call-handling instructions that come back from the service control point, the switch routes the call. (See also signal, CCITT, integrated services digital network, signal transfer point, service control point.)

signal transfer point (STP) – In Signaling System 7, a specialized packet switch that routes network control information among circuit switches and between circuit switches and the network information databases called service control points. (See also Signaling System 7, signal, service control point.)

SMDR (Station Message Detail Recording) – A feature of private branch exchanges and some key systems that logs all outgoing calls. This feature records the originating extension, number dialed, time of connection, call duration and any user-dialed account codes, with no analysis. Many personal computer programs are available to analyze SMDR data.

Software Defined Data Network (SDDN) – AT&T's virtual data network product that takes advantage of ISDN and SS7 to perform high capacity virtual data networking. (see Accunet Switched Digital Service.)

Software Defined Network (SDN) – AT&T's virtual network product. Consists of two versions – Basic for the lower end of the market, and Custom, which is more feature-rich.

Southwestern Bell – One of the seven regional holding companies created at divestiture. It is the holding company for Southwestern Bell Telephone. Southwestern Bell also has other subsidiaries engaged in various communications-related and noncommunications businesses.

special access – Dedicated exchange access service provided by local telephone companies. Interexchange carriers use special access for making nonswitched connections between long distance networks and local private lines leading to customers' premises. Special access is provided for a fixed monthly fee.

special access line (SAL) – The physical circuit installed by a local telephone company to provide a nonswitched connection between a long distance company's point of presence and an end-user location.

special access surcharge – A monthly fee added on to the basic charge for any special access line capable of being connected to the local switched network through a customer-owned PBX or other phone system. The fee is waived if the end-user states in writing that interconnection of a special access line and local switched lines is impossible.

speed calling – See abbreviated dialing.

speed dialing – See abbreviated dialing.

SprintNet – Formerly Telenet. The data transmission arm of Sprint Communications Corp.

Sprint FAX – Sprint's facsimile network product.

Sprint Gateways – Sprint's 900 service bureau that provides all enhanced processing on 900 calls.

SS6 – See signaling system 6.

SS7 – See signaling system 7.

station – 1) A telephone set or location at a business site. 2) Any piece of digital terminal equipment that is capable of receiving or transmitting messages on a data link. See also customer premises equipment.

station message detail recorder (SMDR) – A piece of equipment, generally accompanying a PBX, that can track calls made and received, and generate a record of those calls. Reports can provide call detail and calling totals by individual station or extension. SMDRs are frequently used for monitoring abuse and for charge-backs to company departments.

Subaccount Identification Codes – AT&T's virtual network accounting code option. Nonvalidated three-digit codes available on basic SDN that help users allocate costs among cost centers. They are not used with authorization codes. See also account codes.

Super Access – In cellular communications, a service that allows certain cellular users to receive calls made to their cellular telephone numbers when traveling outside their home service areas.

switch – A major piece of any communications network, a switch is used to interconnect circuits to form a communications path between other switches, PBXs, or other peripheral devices. The process may be implemented using mechanical, analog electronic or digital electronic technologies.

switched access – 1) Exchange access provided by the local telephone carrier that uses the Public Switch Telephone Network (PSTN) in order to deliver the call from the origin point to the termination point. Typically, switched access refers to the delivery of calls from a local subscriber to the long distance carrier point of presence. 2) Describes charges borne by a long distance carrier for the origination and termination of calls in the local exchange area. Switched access fees are per-minute fees. (See also feature group, point-of-presence, special access.)

switched circuit – A connection through the public switched telephone network for which the physical path, established by a switch in response to dialing, is determined by available transmission facilities at the time of the call. As such, the path may change with each call.

switched digital access – See digital switched access.

synchronous transmission – Fixed rate data transmission in which the sending and receiving stations are synchronized by special timing circuits for the duration of the sending of a block of data or for the whole transmission. This timing synchronization allows the sender to eliminate start and stop bits in defining each character. The primary alternative to synchronous transmission is asynchronous transmission, in which the sending and receiving stations are synchronized for only the duration of a single character.

Systems Network Architecture (SNA) – The primary information transmission structure for IBM networks. This includes the layered logical construct, formats, and protocols for direct transmission.

T-1 – A type of digital carrier system transmitting voice or data at 1.544 Mbps. A T-1 carrier can handle up to 24 multiplexed 64 kbps digital voice and data channels. A T-1 carrier system can use metallic cable, microwave radio or optical fiber as transmission media. (See also DS-1, digital hierarchy.)

T-3 – A type of digital carrier system transmitting voice or data at 44.6 Mbps. A T-3 carrier can handle up to 28 multiplexed T-1s or 672 multiplexed 64 kbps digital voice and data channels. T-3 carrier systems use microwave radio or optical fiber as transmission media. (See also DS-3, digital hierarchy.)

tandem data circuit – A data channel passing through two or more data circuit-terminating equipment (DCE) devices in tandem (operating in parallel).

tandem office – A major local exchange carrier switching center for the switched telephone network, which interconnects two or more central offices that cannot directly be connected; a major switching center linking several end offices, especially in high-density areas.

tandem switching system – A network arrangement that enables a call to be routed through two or more switching centers in tandem (operating in parallel).

tariff – The schedule of rates and regulations commonly set by communications carriers and filed with the appropriate federal and state regulatory agencies; the published official list of charges, terms and conditions governing provision of a specific communications service or facility, which functions in lieu of a contract between the subscriber or user and the supplier or carrier.

telco – A generic abbreviation for telephone company. (See also telco (independent), telephone company).

telco, independent – A telephone company that was never a part of the predivestiture Bell system and which remains outside of the postdivestiture regional Bell holding company structure; a telephone company that is not a Bell operating company. (See also telephone company.)

telecommunications – A generic term describing one-way and two-way transmission of information by means of electromagnetic or electro-optical systems, i.e. communications over distance.

telecom – A shortened form of the word telecommunications.

teleconference – A conference between persons linked by a telecommunications system.

telephone company – In this book, a generic term for the entity which provides local and intrastate long distance services to a group of local telephone subscribers. Telephone companies are in contrast to the long distance carriers, which provide intrastate and interstate long distance services.

terminal – A device in a network which is capable of either originating or receiving information conveyed over communication lines. A terminal device may be designed to handle voice transmissions, data transmissions or both. Terminals may be equipped with a telephone handset, a keyboard, a visual display, a printer and more, depending on the intended function.

tie line – A dedicated telephone circuit provided by a common carrier to link telephones at two different premises.

tie trunk – A telephone line that directly connects two private branch exchanges (PBXs).

time division multiple access (TDMA) – A satellite transmission technique which provides multiple, and apparently simultaneous, transmissions to a single transponder with a minimum of interference.

time division multiplexing – The method of deriving multiple channels on a single transmission facility by connecting terminals one at a time in a recurring sequence.

time division switching – A time division multiplexing (TDM) channel-switching method, which switches the output of a particular terminal from one multiplexed channel to another by connecting that terminal at a different time within the recurring sequence of terminal connections. (See also time division multiplexing.)

toll – 1) A service charge for long distance calls. 2) A general term applied to the plant used for toll call services.

toll center – A telephone central office circuit switching facility. It connects end offices to long distance (toll) circuits and can make connections with other central offices to increase routing options.

toll restriction – Method of blocking a telephone user's access to the toll network; permits users to dial local calls, but blocks toll calls or calls to a toll operator without attendant assistance.

tone dial signaling – The electronic signal emitted by the circuitry of touch-tone push-button dials which represents the dialed digit. Tone signals are compatible with all local and long distance phone networks and most computer information services. (See also DTMF.)

touch-tone – See dual-tone multifrequency signalling.

traffic – A generic term that includes any and all calls, messages and data sent and received by means of telecommunications.

traffic engineering – The process of engineering a telecommunications network for maximum efficiency.

transmission – The sending of electrical or light signals on electric wires, radio or optical fiber.

transparency – 1) When two terminals, logical units or devices may communicate with each other unaware of any intervening equipment and software. 2) That property of transmission media which allows transmission of signals without excessive distortion.

transport – That portion of a call cost that pertains only to the carriage of the call by a carrier. This is separated from the billing and collection services that a carrier may charge for.

trunk – A telephone circuit with a switch at both ends. A trunk may connect two central office switches, or two PBXs, or a PBX and a central office switch. (See also line, switch, PBX.)

trunk grouping – A method of reducing the likelihood of traffic blockage due to network congestion. Trunks connecting two switching centers are grouped in a manner such that dialing a single trunk number offers access to the entire trunk group. The call is automatically put on whichever trunk in the group is idle. (See also trunk, switch.)

trunk-side connection – A telephone circuit connected to a central office through a switch port that has been programmed to treat the circuit as a trunk circuit connecting to another central office. Trunk-side connections offer all of the network transmission and signaling features appropriate for a circuit interconnecting two central office switches, and use a separate path for transmission in each direction. (See also trunk, feature groups, access charges.)

unbillable – With 900 services, a call that cannot be billed by the LEC, because the caller moved, canceled the account, etc.

unbundling – 1) The separate pricing of hardware, software and related services. 2) The separate pricing of each component or element of a communications product or service, so that the customer may select only those components or elements he needs without having to accept unnecessary elements or components.

uncollectible – With 900 services, a call that was billed by the LEC but cannot be collected.

uniform call distribution (UCD) – A PBX feature which assigns incoming calls to designated stations; distributing the calls as smoothly and as evenly as possible among stations.

uniform numbering plan – Permits station users at a PBX to place calls over tie trunks using a uniform dialing plan. The numbering plan consists of an access code, a two- or three-digit location code, and a two-to-four-digit telephone code.

Universal Range Privilege – A call screening feature under MCI's Vnet service, a customer can specify the type of Vnet call allowable for each dedicated access line,

Vnet Card authorization code, Dial 1 ANI, and for each ID code. Range options: 1) no calls allowed; 2) on-Vnet Numbers; 3) option 2 plus ten-digit off-Vnet numbers in U.S.; 4) option 3 plus international numbers, or 5) North American Numbering Plan only.

user – 1) Any person, group or organization that employs a telecommunications device, system or network for the transfer of information to others. 2) Any person, group or organization that employs a computer, computer system or computer network to create, process or transfer information.

US Sprint – A major long distance company that competes against AT&T. Formed by the merger of GTE Sprint and U.S. Telecom, US Sprint is the third largest long distance company in terms of market share. It offers nationwide service.

US West – One of the seven regional holding companies resulting from the divestiture of AT&T. US West is the holding company for the Mountain Bell, Northwestern Bell and Pacific Northwest Bell telephone operating companies. US West also owns several other subsidiaries engaged in various lines of communications-related and noncommunications businesses.

video teleconferencing – A two-way telecommunications service that allows "live" moving TV pictures and speech of participants in a conference to be transmitted between two or more locations. Videoconferencing services generally require the digital T-1 (1.544 Mbps) transmission rate or higher transmission rates. (Also called videoconferencing.)

videotex – A form of two-way interactive data transmission.

virtual circuit – A network arrangement which enables information to be transmitted from one location to another over any number of physical circuit configurations during a single period of communication, the whole process controlled by computer software in a fashion imperceptible to the users. Packet switched data networks are a common example of virtual circuits in action. (See also packet switching.)

virtual network services – Switched telecommunications services that offer an alternative to private networks and special service circuits. Traffic is routed over a switched network under the control of network computers. The computers are programmed to provide capabilities that are functionally identical to those offered by special service circuits or networks of private lines. To the users, a virtual network service appears indistinguishable from a hardware-based network. (See also SDN, Vnet, VPN.)

virtual WATS – Discount long distance service for high-volume outward calling, with a pricing structure resembling that of WATS but which employs switched access lines and mileage based banding instead of the dedicated access lines and geographic bands of traditional WATS service. All major long distance companies offer virtual WATS services under registered trade names such as AT&T Pro WATS or MCI Prism. (See also WATS.)

Vnet – MCI's virtual network product.

voice-grade channel – A channel with an analog frequency range between 300 and 3000 Hz or a digital bandwidth of 64 kbps that is typically used for speech transmission, but can also be used for analog or digital data or facsimile transmission.

Vnet 56 kbps – MCI Vnet's switched 56 kbps service. Requires dedicated T-1 facilities.

VPN – Sprint's virtual network product. (See also virtual network services.)

VPN 56 – Sprint's switched 56 kbps data product that is part of its VPN service.

WATS – See Wide Area Telecommunications Service.

WATS Access Line (WAL) – An arrangement which is provided by the local exchange carrier to connect a customer premises with its serving central office through dedicated facilities and then through switched facilities to either a terminating switched access location within the LATA or a long distance carrier's point of presence. A WAL differs from a dedicated access line (DAL) in that the local telephone company can screen traffic coming from the customer premises for intra-LATA calls that the interexchange carrier is not supposed to carry. (See also DAL.)

WAL – See WATS Access Line.

Wide Area Telecommunications Service (WATS) – A bulk rate long distance service providing both outbound and inbound service. WATS services are more of a pricing mechanism than a physical circuit or hardware. Generally, outbound users can call from their location to anywhere in the world; users of inbound (800) service can receive calls from a similarly broad area, although international access is expensive. A customer is connected to the long distance carrier either by equal access, WAL or DAL access, for either form of WATS service. (See also bands, dedicated access line, 800 Service, equal access, WATS access line.)

wire center – The building from which a customer's local service is provided. This includes any access or egress provided for long distance carriers.

wireline – In cellular communications, a term given to the company operating the B block of cellular channels.

World Connect – An AT&T program that gives users extra discounts on customer-dialed calls to all IDDD countries, for all dedicated access SDN users. In 1990, the program was merged with AT&T's Megacom service.

X – In a discussion of numbering plans, X stands for any digit from 0 through 9. For instance, exchange numbers are generically denoted NXX, where N stands for any digit from 2 through 9, representing the numbers 200-999. (See also N, North American Numbering Plan, numbering plan.)

Index

The Author

Daniel D. Briere is president of TeleChoice, Inc., a telecommunications consulting firm in Manchester, CT, that offers custom research and support services to such clients as the major interexchange carriers and a host of other carriers and user. He is currently a contributing editor to *Network World* and *Global Networks,* trade papers of IDG Communications Publications, Inc. He has been managing editor of *Telecom Manager*, which dealt with user issues, and contributing editor of *Intelligent Network News,* which covers developments in intelligent network services and equipment. He holds a Masters of Business Administration from the Fuqua School of Business and a Bachelor of Arts degree from Duke University.

Mr. Briere has published several books and white papers on long distance and intelligent network subject, including *Long Distance Alternatives: T-1, Fiber Optics and Beyond; Virtual Networks: A Guide to the New Generation of Virtual Network Services*; and *Virtual Network: A Buyer's Guide.*

TeleChoice, Inc.
210 Pine Street Suite 115
Manchester, CT 06040
(203) 645-0471

The Artech House Telecommunications Library

Vinton G. Cerf, *Series Editor*

A Bibliography of Telecommunications and Socio-Economic Development by Heather E. Hudson

Advances in Computer Systems Security: 3 volume set, Rein Turn, ed.

Advances in Fiber Optics Communications, Henry F. Taylor, ed.

Broadband LAN Technology by Gary Y. Kim

Codes for Error Control and Synchronization by Djimitri Wiggert

Communication Satellites in the Geostationary Orbit by Donald M. Jansky and Michel C. Jeruchim

Current Advances in LANs, MANs, and ISDN, B.G. Kim, ed.

Design and Prospects for the ISDN by G. DICENET

Digital Cellular Radio by George Calhoun

Digital Image Signal Processing by Friedrich Wahl

Digital Signal Processing by Murat Kunt

E-Mail by Stephen A. Caswell

Expert Systems Applications in Integrated Network Management, E.C. Ericson, L.T. Ericson, and D. Minoli, eds.

Handbook of Satellite Telecommunications and Broadcasting, L. Ya. Kantor, ed.

Innovations in Internetworking, Craig Partridge, ed.

Integrated Services Digital Networks by Anthony M. Rutkowski

International Telecommunications Management by Bruce R. Elbert

Introduction to Satellite Communication by Bruce R. Elbert

International Telecommunication Standards Organizations by Andrew Macpherson

Introduction to Telecommunication Electronics by A.Michael Noll

Introduction to Telephones and Telephone Systems by A. Michael Noll

Jitter in Digital Transmission Systems by Patrick R. Trischitta and Eve L. Varma

LANs to WANs: Network Management in the 1990s by Nathan J. Muller and Robert P. Davidson

Long Distance Services: A Buyer's Guide by Daniel D. Briere

Manager's Guide to CENTREX by John R. Abrahams

Mathematical Methods of Information Transmission by K. Arbenz and J.C. Martin

Measurement of Optical Fibers and Devices by G. Cancellieri and U. Ravaioli

Meteor Burst Communication by Jacob Z. Schanker

Minimum Risk Strategy for Acquiring Communications Equipment and Services by Nathan J. Muller

Mobile Information Systems by John Walker

Optical Fiber Transmission Systems by Siegried Geckeler

Optimization of Digital Transmission Systems by K. Trondle and G. Soder

Principles of Secure Communication Systems by Don J. Torrieri

Principles of Signals and Systems: Deterministic Signals by B. Picinbono

Private Telecommunication Networks by Bruce Elbert

Quality Measures and the Design of Telecommunications Systems by John H. Fennick

Radiodetermination Satellite Services and Standards by Martin Rothblatt

Setting Global Telecommunication Standards: The Stakes, The Players, and The Process by Gerd Wallenstein

Signal Theory and Processing by Frederic de Coulon

Techniques in Data Communications by Ralph Glasgal

Telecommunications: An Interdisciplinary Text, Leonard Lewin, ed.

Telecommunications in the U.S.: Trends and Policies, Leonard Lewin, ed.

Telecommunication Systems by Pierre-Girard Fontolliet

Television Technology: Fundamentals and Future Prospects by A. Michael Noll

Terrestrial Digital Microwave Communications, Ferdo Ivanek, ed.

The ITU in a Changing World by George A. Codding, Jr. and Anthony M. Rutkowski

The Law and Regulation of International Space Communication by Harold M. White, Jr. and Rita Lauria White

The Manager's Guide to the New Telecommunications Network by Lawrence D. Gasman

The Telecommunications Deregulation Sourcebook, Stuart N. Brotman, ed.

Traffic Flow in Switching Systems by G. Hebuterne

Troposcatter Radio Links by G. Roda

Voice Teletraffic System Engineering by James R. Boucher

World Atlas of Satellites, Donald M. Jansky, ed.